The Research Report on Application of Low-carbon Technology in Expo 2010 Shanghai

UNIDO International Solar Energy Center
for Technology Promotion and Transfer
(UNIDO-ISEC) · Jiangsu Modern Low-carbon
Technology Research Institute
Editors

The Research Report on Application of Low-carbon Technology in Expo 2010 Shanghai

SSTP

Springer

Editors
UNIDO International Solar Energy Center for Technology Promotion and Transfer (UNIDO-ISEC)
Lanzhou
China

Jiangsu Modern Low-carbon Technology Research Institute
Jiangsu
China

ISBN 978-3-662-51317-0 ISBN 978-3-662-44357-6 (eBook)
DOI 10.1007/978-3-662-44357-6

Springer Heidelberg New York Dordrecht London

Jointly published with Shanghai Scientific and Technical Publishers, Shanghai
ISBN: 978-7-5478-1023-1, Shanghai Scientific and Technical Publishers, Shanghai

Printed on acid-free paper

Springer is part of Springer Science+Business Media (www.springer.com)

Editorial Board

Producers	Xi Wenhua, Xu Yongjun
Expert Review Board	Shi Dinghuan, Xu Dingming, Zhang Ming, Dai Shunzhi, Qian Zhixin, An Fengquan, Yang Hongwei, Wu Changhua, Jin Baosheng, Huang Qiang, Qian Gang, George Hamilton (USA), Jacqueline Faulkner (UK)
Translators of English	Sun Bingli, Xi Huichao
Proofreaders of English	Chen Dongling, Chalmers Benedict Wood (USA)
Photographer	Hu Haisong

Congratulatory Message from Dr. Kandeh K. Yumkella, Director-General of UNIDO

The Shanghai EXPO, which attracted worldwide attention, met its goal of being a "low-carbon EXPO." Through a series of low-carbon designs, low-carbon constructions, low-carbon technologies, and low-carbon operational procedures, visitors were able to experience the low-carbon concept that existed in every corner of the EXPO Park.

This research report is a summary of the energy-saving, low-carbon, and environmentally friendly technologies that were demonstrated in over 200 pavilions in the EXPO Park. As such, it shows the wealth of ideas which humanity has come up with to make our societies low-carbon and environmentally friendly.

I am sure this report will effectively contribute to our common efforts to adopt low-carbon, energy-saving, and emission-reducing technologies. In so doing, it will help promote a global development that is economically, socially, and environmentally sustainable. It is therefore conducive to the realization of the Millennium Development Goals, to the elimination of poverty, and to the improvement of the quality of life of people in general, in particular in the developing countries.

On behalf of UNIDO, I extend my warmest congratulations to Prof. XI Wenhua and all the participants and experts who were involved, for their hard work in researching, compiling, and publishing this report.

Dr. Kandeh K. Yumkella
Director-General, UNIDO

Report of Low-carbon Technologies in the Post-Expo Era (Popular Edition)

—Origin and Background of *The Research Report on Application of Low-carbon Technology in Expo 2010 Shanghai*

When the Shanghai Expo was enshrined in history with applause, the world may remain overwhelmed by the gorgeous pavilion exhibition that somewhat overshadows extensive application of new low-carbon technologies in the Expo that are energy efficient and environmentally-friendly.

The Chinese premiere Wen Jiabao stated in his speech while addressing the Expo Summit: "The Shanghai Expo has championed new development concepts such as green and low-carbon growth. The Expo Park itself is a paragon of low carbon application with numerous pavilions built with the latest carbon efficient materials and energy saving technologies; the low-carbon technologies and products on display are innumerable." From April to October 2010, I moderated or hosted a total of five forums and exhibition activities in the UN Pavilion, over which I found the Expo not only a historical, cultural, artistic, and social demonstration of various countries, but also an encouraging showcase of the world's leading technologies featuring energy-saving, green, and low-carbon techniques, hence I carefully observed myriad engineering products of these technologies within very short span of time. Admittedly, they are fascinating.

It is evident that state-of-art technologies, particularly sophisticated energy-efficient, green, and low-carbon ones were exhibited to varied degrees in over 200 pavilions. I am soberly aware that these are technologies we need to mitigate global climate change and catalyze low carbon economies.

However, how much do Expo visitors know about these technologies? How to continue using these technologies in the post-Expo era? How to disseminate them in a wider community? These are lingering questions I reflect upon and what reassures my commitment to preserving these low-carbon technologies within the public scene.

It is undoubtedly hard to collect, study, and summarize these applications after the Expo when the bulk of pavilions will be demolished. We have to, therefore, race against time to survey, analyze, and test them over the Expo and consolidate the result in a research report. I started my initiative right after hosting the Low Carbon Technology Shanghai Action Forum in the UN Pavilion on May 4 of 2010.

I searched for proper partners everywhere; many flinched as the job is tough. It was not long, however, before Mr. Xu Yongjun, Dean of Jiangsu Modern Low-Carbon Technology Research Institute reached a consensus with me. We finally entrusted Jiangsu Modern Low-Carbon Technology Research Institute in an organization of over 30 experts and postgraduates from Nanjing University and Nanjing University of Technology to make a comprehensive survey of approximately 200 pavilions in the Shanghai Expo. Our team eventually completed this *Research Report on Application of Low-Carbon Technology in Expo 2010 Shanghai* within 2 months, leaving a precious heritage of the Expo for the global progress in low-carbon economic development.

The Research Report on Application of Low-Carbon Technology in Expo 2010 Shanghai echoed four needs for development: first, all government agencies expect to learn environmental-friendly technologies such as low-carbon and energy-saving technologies to economize energy and reduce emission for mitigation of climate change; second, entrepreneurs are seeking for advanced technologies through the Shanghai Expo to invest for profits; third, research institutions and universities expect to know the world's level of energy conservation and emission reduction through the Expo for R&D breakthroughs; fourth, all business and organization exhibitors in the expo, especially those offering new products and technologies, such as Chinese solar power enterprises including Changzhou Trina Solar Limited., Micoe Solar Energy Technology Co., Ltd., Beijing Tianpu Solar Power Industry Co., Ltd., Jiangsu Sunrain Solar Energy Co., Ltd., Nanjing Nanzhou New Energy Co., Ltd., and Shanghai Ebays Energy Saving Technology Co., Ltd. need such a report to summarize and recognize their technological achievements for broader application.

On Oct. 31, 2010, the Assessment Conference on *Post-Expo Low Carbon New Era Forum—the Research Report on Application of Low-Carbon Technology in Expo 2010 Shanghai* was held in Changzhou city of Jiangsu Province, whereby experts and scholars from research institutions, universities and government organs like National Energy Bureau, National Energy Experts Consulting Committee, China Renewable Energy Society, China Architecture Science

Research Institute, Gansu Natural Energy Research Institute, Institute for Sustainable Communities (ISC), UK-based Zero Carbon Centre (ZED), Nanjing University, Nanjing University of Technology, Jiangsu Provincial Technology Department, Jiangsu Provincial Development and Reform Commission highly acknowledged and recognized the research report, which fuelled our confidence to expedite its publication.

The Chinese president Hu Jintao stressed in his moderation of intensive study program for Political Bureau of the CPC Central Committee on Dec. 3, 2010: "We must seriously absorb and apply the latest technological achievements exhibited in Shanghai Expo; actively study world technology development trend; try to achieve more breakthroughs in key technologies and obtain more intellectual property rights; actively nurture and develop strategically important emerging industries; speed up training of creative talents to build an innovative country." Our act of making this report is exactly such an effort. We hope its publication could make some contribution to China and to the world's transition to sustainable development with climate change mitigation and low carbon economy achieved via scientific breakthroughs, which is both our intention and motive for persistent work.

The research report is published both in Chinese and English for local and international readers. It cannot be made possible without support and encouragement from all colleagues, friends, agencies, and organizations home and abroad including China National Energy Bureau, Ministry of Science and Technology, Ministry of Commerce, Ministry of Industry and Information Technology, Foreign Affairs Commission of CPPCC, active cooperation by exhibitor pavilions of countries and regions and in particular, strong commitment from Director Su Jianghua of Jiangsu Wujin Hi-Tech Industrial Zone Management Committee and management of Shanghai Scientific and Technical Publisher as well as editor of this report, for which I hereby would like to extend my earnest gratitude. Given the breadth of disciplinary areas the report covers and the limit of time for this first edition, errors and oversight may be unavoidable, hence correction and comments from our readers are much appreciated for better editions that will follow.

January 2011

Prof. XI Wenhua
Director-General
UNIDO International Solar Energy Center for
Technology Promotion and Transfer (ISEC)
In Cuidao Island, Guangzhou, China

Contents

Abstract

As a research report, this book focuses on the low-carbon technologies shown in Expo 2010 Shanghai, China, which covers the utilization and application of renewable energy, new type of low-carbon technologies, low-carbon construction, water treatment, waste disposal, low-carbon transportation, etc. It draws together and analyzes data collected from the Expo site on several aspects from the initial planning and design, pavilion construction, to operation management, concept demonstration and some enterprises involved, with a summary at the end of each section. The book comprises five chapters, which starts with an overview of the World Expo. Chapter 2 discusses and reviews the application of low-carbon technologies in the Expo. Chapters 3 and 4 look at the typical pavilions with low-carbon characteristics and low-carbon life concept. Recommendations are made in the final chapter. The appendices of the book offer a list of companies and organizations with their products and other relevant information. These companies and organizations illustrated in the Expo the technologies of new and renewable energy, energy saving in construction, water treatment and new energy vehicles, etc. They provided full support to the Expo with their technologies, products, and services.

This book should be useful and helpful to the professionals relevant to low-carbon economy and environmental protection enthusiasts, as well as ordinary people who want to know low-carbon knowledge. The authors of the book hope that the people in the world longing for an even better urban life would give their concern and supports to the future development of low-carbon technologies.

Abstract

Chapter 1
General Survey of the World Expo

In the long process of human history, the World Expo has not only demonstrated the fruitful achievements of material civilization of its times, but also reflected people's understanding of industry, technology, and culture. Taken in this sense, the history of the World Expo is also a history of people's evolving and progressing evaluation of the civilization they created.

To some extent, the Industrial Revolution of Europe gave birth to the modern World Expo, as the original intention of the Expo was to showcase the results of the industrial revolution of different countries. The first World Expo was held in London, UK in 1851, under the title "the Great Exhibition of the Works of Industry of All Nations" or "the Great Exhibition." Lasting for 160 days, the exhibition invited over 10 countries to participate, attracting more than 6.3 million visitors. The Great Exhibition took place in Hyde Park in the downtown area of London. The exhibition building was built of iron and steel frames and glass curtains, so it was called "the Crystal Palace."

The 1900 Exposition Universelle in Paris inherited the technology-centered tradition of the Expo and displayed the technological achievements of the nineteenth century from the Western world. Visitors to the Expo unprecedentedly reached 48.1 million.

In 1933, the United States held the Chicago Expo. Forty-seven countries participated in the exhibition and 38.3 million visitors attended the event. The Chicago Expo featured two creations. Firstly, for the first time the Expo set a specific theme —*A Century of Progress*. Henceforth, an officially declared theme became the core of every Expo. Secondly, it created an era of large corporation building pavilions in the Expo site. The Expo allowed large industrial corporations such as General Motors, Ford, and Chrysler to build their dedicated pavilions, a pattern welcomed by both entrepreneurs and visitors and taken as the practice in the Expos afterwards.

The 1958 Brussels Expo was the first Expo after World War II, which was held at a time when people were not only rebuilding their homes from the ashes of war but also restoring their faith in progress. The theme of this Expo was "A world view —A new humanism," which symbolized the conversion of emphasis from

UNIDO-ISEC and Jiangsu Modern Low-carbon Technolgy Research Institute (eds.),
The Research Report on Application of Low-carbon Technology in Expo 2010 Shanghai,
DOI 10.1007/978-3-662-44357-6_1

technology-centered ideas to humanism. This shift is reflected in the landmark of this Expo—the Atomium, which was a giant model of the atomic structure, symbolic of man's use of atomic energy in a safe and peaceful way.

The World Expo of Osaka in 1970 attracted 76 countries and four international organizations to participate in and 64 million visitors to attend the event. This Expo gave a strong boost to the development of local transportation, business facilities, tourism, and cultural exchange facilities in Osaka, promoted the rise of a city cluster of the Kansai region with Osaka as the core, which was of great significance to the development of Japan's national economy. The Osaka World Expo was also the first World Expo held by an Asian country.

From the twentieth to the twenty-first century, people had experienced a transition from conquering nature to respect for nature. Apart from focusing on the development of the host city itself, the World Expo began to take the whole earth in view and much concern was shown to global issues such as ecology and environmental protection. Meanwhile, the numerous wars of the past century filled people with a strong longing for peace. So the World Expo was viewed by people around the world as a bridge for communication to eliminate misunderstanding, enhance mutual understanding, and resolve contradictions.

Inspired by the above ideas, the organizer of the World Expo Hanover in 2000 decided to choose "Protecting the Resource" as one of its topics, which was in line with the trend of the times. The 2005 Aichi Expo, Japan, the most recently registered Expo, was themed "Nature's Wisdom," which presented a global perspective, reflecting man's attitude toward nature, society, and future. The Expo Zaragoza 2008, the latest recognized Expo, was an international event with the theme around the dynamic relationship between water and human society; it was a global supportive program with profound impact based on the lasting concept of water and sustainable development.

The World Expo in the twenty-first century is still full of vitality, and will continue to be the impetus to push forward the development of technology and civilization. Since the Industrial Revolution, people have started to use nonrenewable fossil fuel at large scale with consumption increasing at 2 % per year, which results in rapid increase in carbon dioxide emissions followed by ecological imbalances such as natural disasters, global warming, and energy crisis. "Low-carbon" has inevitably become a global hotspot.

Under the theme of "Better City, Better Life", Expo 2010, Shanghai, China (the Shanghai Expo) fully reflects people's longing for better cities and a better life around the world. What is particularly noticeable about the event is the demonstration of the latest low-carbon technologies and ideas, as well as low-carbon cities and lifestyles by more than 200 countries, international organizations, and enterprises. Under the guidance of UNIDO International Solar Energy Center (ISEC-UNIDO), Jiangsu Modern Low-carbon Technology Research Institute keenly seized the opportunity of the Expo and organized more than 30 professors and graduate students engaged in a 2-month intensive field survey on the application of low-carbon technologies in the Expo, from the overall design and planning of the Expo site to its operation and management, from concepts demonstration to technologies application enterprises.

The low-carbon technologies involved in the above survey include: development and application of renewable energy (solar, wind, biomass, fuel cells, etc.), new low-carbon technologies (ground-source heat pump, water pump, ice storage technology, non-electric air conditioning, smart grid, LED lights, etc.), low-carbon buildings (energy-saving materials, energy-saving technologies), water treatment (rain water recycling, water treatment), and low-carbon transportations.

We hereby dedicate this book to Expo 2010 Shanghai, China, and expect greater attention to be paid to low-carbon cities and low-carbon lifestyles as well.

Chapter 2
Application and Analysis of Low-Carbon Technologies in Expo 2010 Shanghai

The use of new materials that are energy-efficient and eco-friendly has become one of the highlights of Expo 2010 Shanghai, China, as seen in the large solar panels on pavilion roofs, the LED lights illuminating the whole Expo Park, and the new energy vehicles shuttling between pavilions. This is an initiative to put into practice the low-carbon concept and to test and demonstrate how the new energy is to be used in the future. Undoubtedly, the successful integration of "green" and "low-carbon" elements has better embodied the theme of Expo 2010 Shanghai China, "Better City, Better Life."

In the Expo Park, the solar power technology, wind turbine, new energy-driven vehicles and smart power grid (SPG), and other technologies were extensively used, most of which will be popularized after the Expo. Expo 2010 has been so far the largest demonstration area for new energy-driven vehicles to enter into operation. Over 1,000 new energy-driven vehicles running in the Expo site led to reduction of 13,000 tons of carbon dioxide emission, ensuring that public transport inside the Expo achieved "zero emission." As for the solar power photovoltaic (PV) generation, the building-integrated solar PV system (BIPV) designed by the Expo organizer had the largest applications with the most technologies involved. A solar power generation system of 4.68 MW was installed on the roofs and glass curtain walls of the China Pavilion, the Theme Pavilions, the Expo Center, and the Art Performance Center, which can reduce 4,000 tons of carbon dioxide emission annually and has been the largest solar PV cells demonstration area in China. The "power generated floor" in Japan Pavilion can transform the energy treaded by humans into electric energy. The solar cells on the outer wall of the Japan Pavilion enabled the wall to generate energy by itself. All the construction materials used in the UK Pavilion could be recycled, with zero carbon dioxide emission in the whole building. The curtain outside the Switzerland Pavilion was made from degradable soybean fiber, which is capable of generating power that can be degraded naturally. The roof of the pavilion was covered with grass and dandelion. Once visitors take the cable car to the rooftop, they can smell the fragrance of the dandelion.

UNIDO-ISEC and Jiangsu Modern Low-carbon Technolgy Research Institute (eds.),
The Research Report on Application of Low-carbon Technology in Expo 2010 Shanghai,
DOI 10.1007/978-3-662-44357-6_2

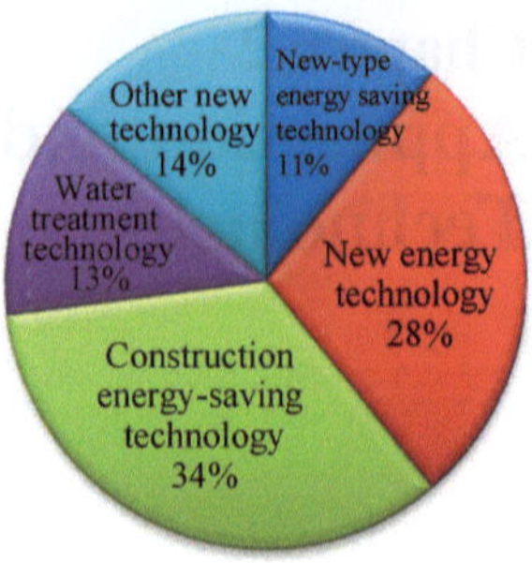

Fig. 2.1 Proportion of different low-carbon technologies applied in the Expo

In terms of the green lighting application, LED lights were used in China Pavilion, Expo Axis, Expo Performance Center, and Urban Best Practice Area (UBPA). This was the first time LED lighting was being used so extensively in city blocks in China.

The Shanghai Expo also served as a large platform for showcasing the new energy technology, energy conservation technologies, water treatment technologies, low-carbon transportation technologies, waste disposal technologies, etc. Figure 2.1 shows a proportion of the various low-carbon technologies applied at the Expo site.

2.1 Renewable Energy and New Energy

Renewable energy refers to energies that are renewable and replenished regularly or recycled, including solar energy, wind power, water power, biomass, tidal energy, etc. Most renewable energies are in fact storehouses of solar energy.

2.1.1 Solar Energy

2.1.1.1 Overview

Solar energy is a radiant energy that comes from a celestial body (the Sun) outside the Earth. A majority of energies that people use come directly or indirectly from the Sun. Various kinds of plants convert solar energy into chemical energy through photosynthesis so that it can be stored in the plants. Fossil fuels such as coal, oil, and natural gas are formed by plants and animals buried underground over long geological ages. They are in essence the solar energy fixed from ancient creatures. Besides, hydro energy, wind energy, wave energy, and ocean current energy are all converted indirectly from solar energy.

Fig. 2.2 Solar water heaters

(1) **Major Uses of Solar Energy**

(i) *Solar Energy Photo-thermal Utilization*

The principle of photo-thermal utilization is to first collect the solar radiation energy through its interaction with a substance and then convert it into heat energy for utilization. Currently, there are three types of solar thermal collectors that are mostly used:flat plate collectors, evacuated tube collectors, and concentrating collectors. Based on the different temperature and different usage, solar photo-thermal utilization is normally classified into low-temperature utilization (<200 °C), medium-temperature utilization (200–800 °C), and high-temperature utilization (>800 °C).

Apart from solar water heaters (Fig. 2.2), solar photo-thermal applications include solar houses, solar cooker, solar greenhouses, solar drying system, solar earth ozone sterilization technologies, etc.

(ii) *Solar PV Power Generation*

Solar power generation is an effective way of making use of solar energy. There are a number of methods for solar power generation, of which the following two have already been put into practical use.

a. Light-heat-electricity conversion. It utilizes thermal energy produced by solar radiation. First, it uses the solar thermal collector to convert the absorbed heat into steam, and then generates electricity driven by the steam turbine. The former process is the conversion from light to heat and the latter is from heat to electricity.
b. Light-heat conversion. The basic principle is to transfer solar radiation energy directly to electrical energy by making use of PV effect. Its key device is the solar cell. Figure 2.3 shows the solar PVpanel.

(iii) *Photochemical Utilization*

It is a light-chemical conversion using the sun's radiant energy to decompose water directly into hydrogen.

Fig. 2.3 Solar PV panel

(iv) *Photobiological Utilization*

It is a process of converting solar energy into biomass through the photosynthesis of plants. At present, these include fast-growing plants such as firewood forest, oil crops, giant seaweed, etc.

(2) **Development and Utilization of Solar Energy**

People have long been working at how to make use of solar energy. But only a small portion of the energy radiated by the Sun strikes the Earth, say one part in 2 billion. Yet this amount of energy is enormous, 30,000–40,000 times as much as the total amount of energy needed globally. So it is inexhaustible. Moreover, different from mineral fuels such as oil and coal, solar energy will not lead to greenhouse effect or global climate change, nor will it pollute the environment. For this reason, countries are paying more attention to the utilization of solar energy and competing to develop various new photoelectric technologies and materials so as to extend the application field of solar energy. Especially in the past decade, faced with the two major crises of increasingly dried up exploitable oil and deteriorating eco-environment, people are increasingly longing for the arrival of the "Solar Era." Solar energy is now extensively used, from power generation, heating and water supply to a variety of solar power plants. In some fields, solar energy has entered into the application stage.

Recently, a Russian company incorporated its self-developed solar water-jet propeller and cold spraying propeller into solar pond engineering, and installed an ice groove to the solar pond, thus designing a new type of solar pond suitable for farming. With this design, building a solar pond of 70 m^2, a household of six to eight in 100 m^2 of housing will be supplied with abundant electricity consumption all the year round. Another research institute put forward an idea of a combined solar pond design that integrates solar energy, geothermy, and waste heat in the house by making use of the technology of heat pump and heat pipes, which greatly reduce the cost of solar power generation. In the Ciscaucasian area, solar generation stations can even compete with thermal power stations. Solar energy is available all the year round for air conditioning in summer and heating in winter.

(3) **Advantages and Disadvantages**

(i) *Advantages:*

① Available to all: As the Sun clasps the Earth without regional restrictions, whether land or sea, mountains or islands, solar energy can be directly exploited and utilized and there is no need for mining or transportation;

② Pollution-free: Development and utilization of solar energy does not pollute the environment. It is one of the cleanest energies, which is extremely valuable today when environment pollution is becoming more and more serious;

③ Enormous: Every year the Sun's radiant energy arriving at the surface of the Earth amounts to about 130 trillion tons of coal, which is the largest among the world's exploitable energies;

④ Long-lasting: According to the present rate at which the Sun produces nuclear power, hydrogen reserves will be adequate to maintain for up to 10 billion years, while its lifetime is about several billion years. In this sense, we can say that the Sun's energy is inexhaustible.

(ii) *Disadvantages:*

Dispersity: Although the amount of solar radiation that arrives at the surface of the earth is huge, the energy flux density is very low. On average, on a fine summer day near the Tropic of Cancer, solar radiation has the largest irradiation at noon: the solar energy received at 1 m^2 area perpendicular to sunlight is about 1,000 W; allocated to every day and night of a year, it is only about 200 W. However, in winter it is roughly half the amount, on cloudy days only 1/5, when such energy flux density is very low. Therefore, if you wish to get certain converted power using solar energy, often quite a big set of collecting and conversion equipment is needed, the cost of which is rather high;

Instability: Restricted by natural conditions such as day and night, seasons, geographic latitude and altitudes, and subject to the random factors such as sun, cloud, or rain, solar irradiation in certain areas of the Earth is not continuous or stable, which adds much difficulty for application of solar energy at large scale. In order to make solar energy a continuous and stable energy as well as help it eventually become an alternative energy source that can compete with conventional energy, the problem of energy storage should be effectively solved, i.e., to store the solar radiant energy on sunny days as much as possible for use at nights or on cloudy and rainy days. Yet, currently energy storage is one of the hardest nuts in the use of solar energy;

Inefficiency and high cost: At present the development of solar energy utilization is theoretically feasible in some aspects and is mature in technology. However, due to its low efficiency and high cost, the economical efficiency cannot compete with that of conventional energy resources. For quite a period of time in the future, the further development of the solar energy utilization will be restricted.

(4) **Economic Analysis**

An increasing number of countries are aware that a society of sustainable development should be one that can meet its social needs without endangering its future generation. Therefore, utilizing clean energy as much as possible instead of high carbon mineral energies should become the basic principle for energy construction. With the change in energy forms, conventional energy reserves are declining; as a result, its prices will inevitably rise and environmental pollution control will require more investment consequently.

China is the world's largest producer and consumer of coal. Coal accounts for 76 % of commercial energy consumption, which has become the major source of air pollution in our country. Vigorously developing new energies and using renewable energy technologies will be an important measure to reduce environmental pollution. The energy problem is a global problem and a new energy period will come sooner or later. In the long run, wide applications of solar energy utilization technologies and devices will inevitably constrain the rise in prices of fossil energy.

(5) **The Prospect of the Industry in China**

The promulgation and implementation of China's Renewable Energy Law has provided policy support for the development of the solar energy industry. The signing of the Kyoto Protocol, the unveiling of the environmental protection policy, and the commitment to the international community has brought opportunities to the solar energy industry. The West Development Program has provided the industry with an enormous domestic market. At the same time, the rise in crude oil prices and adjustment of China's energy strategy urge the government to strengthen the support for renewable energy development. All of these will bring great opportunity for China's solar energy industry.

2.1.1.2 Solar Energy Application in the Expo

The solar energy technology used in the Expo accounted for over 50 % of the low-carbon and energy conservation technologies utilization, which was the best demonstration of "Better City, Better Life."

The solar PV energy technology was applied at large scale in Expo 2010 Shanghai, accounting for 87 % of the entire solar energy utilization in the Expo (Fig. 2.4). Table 2.1 gives the use of solar energy in the Expo.

2.1.1.3 Application of Solar PV Technology in Expo 2010 Shanghai

(1) **Power Generation by Solar PV**

In the major pavilions such as China Pavilion, Theme Pavilion, Expo Center, South City power plants as well as the self-built Pavilions of Japan, Australia and some other countries, solar cells were installed on the rooftops and glass curtain

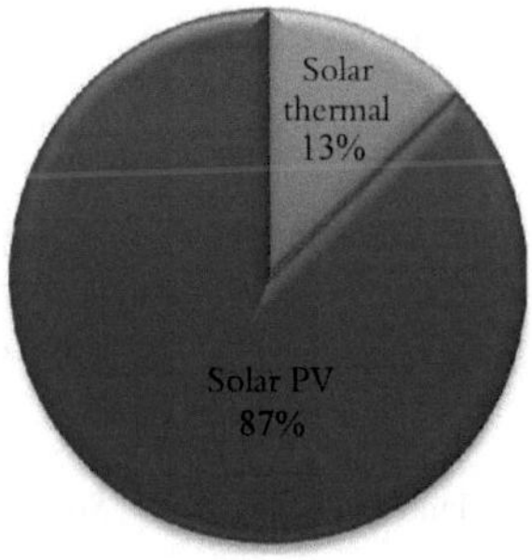

Fig. 2.4 Application of solar energy in the Expo

Table 2.1 List of the solar energy application items in the Expo

Pavilions	Solar energy technology	Applications
Japan Pavilion	Solar photovoltaic cells	Power-generating floor
		Solar panel with membrane technology
Israel National Pavilion	Solar photovoltaic cells	Solar power system
		Pest control device with solar black light technology
India Pavilion	Solar photovoltaic	Solar photovoltaic panels
Expo Axis	Solar energy technology	PECVD filming technology
China Pavilion	Solar photovoltaic cells	Solar photovoltaic power generation system
Macau Pavilion	Solar photovoltaic cells	Solar photovoltaic power generation
Expo Center	Solar photovoltaic	Solar photovoltaic power generation system
World Meteorology Pavilion	Solar photovoltaic cells	Photovoltaic power generation and vehicle recharging column
DEVNET Pavilion	Solar photovoltaic cells	Solar street lights
Theme Pavilion	Solar water heating	New solar water heater
	Solar photovoltaic cells	Megawatt level solar power station
Germany Pavilion	Solar photovoltaic cells	Building integrated solar photovoltaic components
Belgium-European Union Pavilion	Solar photovoltaic cells	Sun-dance flowers
		Solar vehicles
Norway Pavilion	Solar energy	Water purification device with solar energy
Portugal Pavilion	Solar energy	Multifunction solar system

(continued)

Table 2.1 (continued)

Pavilions	Solar energy technology	Applications
Switzerland Pavilion	Solar photovoltaic cells	Solar energy cells
Italy Pavilion	Solar energy	Photovoltaic solar integrated modules that can absorb solar energy, improve the indoor temperature, and power for the pavilions
Finland Pavilion	Solar photovoltaic power generation	Providing electricity for refrigeration equipment of the pavilion
Netherlands Pavilion	Solar photovoltaic power generation	Umbrella affixed with solar panels
		Solar car
USA Pavilion	Solar photovoltaic power generation	Solar photovoltaic power generation
State Grid Pavilion	Solar photovoltaic power generation	Solar photovoltaic panels
Shanghai Enterprise Joint Pavilion	Solar water heating	Generating power with the hot water of 90°
China Private Enterprises Pavilion	Solar photovoltaic power generation	Solar photovoltaic power generation and solar water heater
Information and Communication Pavilion	Solar energy	Solar energy power generation
City Future Pavilion	Solar photovoltaic power generation	Solar energy power generation
Shanghai's UBPA Case	Solar photovoltaic cells	Solar film photovoltaic power generation
	Solar water heating	Photovoltaic solar
Osaka Case Pavilion	Solar photovoltaic cells	Thermoelectric conversion
Madrid Case Pavilion	Solar cooling	50 translucent film solar modules were installed on the rooftop, which can keep any room cool by use of the cooling device, without affecting the building lighting
Hamburg House Case Pavilion	Solar photovoltaic cells	Solar photovoltaic power generation
London Case Pavilion	Solar optothermoel	Solar energy heater
	Solar photovoltaic cells	Wind-solar hybrid public lighting systems
	Solar cooling	Air conditioning refrigeration

(continued)

Table 2.1 (continued)

Pavilions	Solar energy technology	Applications
Alsace Case Pavilion	Solar photovoltaic power generation	Exterior wall solar panel technology
Mobile Communications Base	Solar photovoltaic power generation	"Wind-solar hybrid system" can provide nearly 30 % of the energy consumption of the base station.
Expo Park	Solar photovoltaic lighting	Decorative lights with solar energy
Mexico Pavilion	Solar photovoltaic power generation	Lighting for indoor pavilions
United Nations Pavilion	Solar thermoelectricity	Building integrated with solar and thermoelectricity

walls with a total installed capacity of more than 4.68 MW, which combined with the main grid of Shanghai to generate electricity. Among them, the total installed capacity of the Theme Pavilion was about 2.825 MW, with the solar panels covering 30,000 m^2, and the annual generating capacity reaching about 2.5 million kWh. It has been the largest single solar PV generating project as well as the BIPV project in China. Figure 2.5 shows the solar PV panels on the roof of the Expo Center.

(2) **Solar refrigeration**

The most typical use of solar refrigeration in the Expo was in London's UBPA Case and Madrid's UBPA Case Pavilion.

Solar thermal-power transducers on the roof of London's UBPA Case Pavilion collected enough power for daily use in the building (Fig. 2.6); the collected heat was changed into power for the building refrigeration. According to the survey, by use of various energy-saving measures, the solar power generated output in the London Case Pavilion actually exceeded the needs of the entire pavilion.

On the rooftop of Madrid Pavilion were installed 50 translucent film solar modules (Fig. 2.7), which could cool any room through the refrigeration device, without affecting the lighting in the building.

(3) **Solar PV lighting**

Outside the Swiss National Pavilion, a 17-m-high translucent curtain was trimmed with 10,000 red "Little Suns," as shown in Fig. 2.8. Each "Little Sun" was a disk-sized red photosensitive device making use of the most advanced solar power technology. With the interaction of solar cells, light sensors, and LED, it absorbed

Fig. 2.5 Solar PV panels on the roof of the Expo center

Fig. 2.6 Solar energy systems on the roof of the London's UBPA case

Fig. 2.7 Solar energy system in Madrid Pavilion

and stored light energy and converted it into electricity based on the angle and intensity of light. The "little Suns" shone independently while interacting with the curtains.

Fig. 2.8 "Little suns" of the Swiss National Pavilion

Fig. 2.9 Solar photo-thermal devices on the roof of Shanghai's UBPA case

(4) **Solar Photothermal**

Shanghai is recognized as a model city for building energy-saving in China by the World Wildlife Fund (WWF); "Shanghai's Eco-Home" at the UBPA is just the integral demonstration of the building energy efficiency technologies. On this building, the BIPV film solar PV generating system installed on the southern exterior walls can generate clean power of 11 kWh (Fig. 2.9), and meanwhile acts as the sunshading frame over the roof garden to reduce direct radiant heat. The flat plate solar water heating system on the southern sloping roof integrated with the PV generating system on the roof can produce an average of 900 L of hot water per day, accounting for about 50 % of the total hot water consumption of the pavilion.

(5) **Solar water heating**

On the roof of Shanghai Enterprise Joint Pavilion, there was a solar collector panel of 2,200 m^2, which collected 95 L hot water generated by solar energy. Through the ultra-low temperature power generation technology, the output power

Fig. 2.10 Integrated solar thermoelectric system on the top of the waiting area of the UN Pavilion

can be over 200 kW. This technology opens up new ways of solar power generation; the generated power can satisfy the needs for construction exhibitions and daily power supply in the pavilion.

(6) **Solar PV/photo-thermal building integration**

Solar thermoelectric integration used in the waiting area of the United Nations Pavilion (Fig. 2.10) was developed and designed by Nanjing Nanzhou New Energy Development Co, Ltd. This area, 40 m long, 5 m wide, and 3.8 m high covered an area of 232 m^2. On the roof of this area were laid 150 new type tiles (Hanwa) that generated power and heat at a total installed capacity of 15.12 kW. It could generate electricity of 60–80 kWh after being exposed to sunshine 5 h on a sunny day.

The solar thermoelectric integration in the waiting area of the UN Pavilion designed by Nanjing Nanzhou New Energy Research and Develop Co., Ltd. was the only demonstration in the Expo site, as well as the only one with such a large area.

It was therefore recommended that "solar PV roof" and "photo-thermal wall project" be widely promoted in urban constructions, which can save materials and generate multi-energy, resulting in reduced comprehensive costs. This technology can meet the need of power supply, thermal heating, cooling, as well as hot water for daily use.

For extensive promotion of this technology there are still some difficulties faced: (1) Lack of interworking between the solar industry and the construction industry; (2) Lack of normalization and standardization of the heat-collector components; (3) New building materials for the solar energy house need to be further developed.

(7) **Solar vehicles**

The venues in the Shanghai Expo that displayed solar vehicles were Belgium-EU Pavilion (Fig. 2.11) and Netherlands Pavilion (Fig. 2.12). Belgium-EU Pavilion displayed a new type of solar vehicle. Its body was made of pre-impregnated carbon

Fig. 2.11 Solar vehicle shown in Belgium-EU Pavilion

Fig. 2.12 Solar vehicle shown in the Netherlands Pavilion

composite material, its chassis, torsion box, and scroll bar of integrated carbon fiber hard shell, and its tires were low rolling resistance tires specialized for solar vehicles developed by Michelin. The length, width, and height of the car were, respectively, 4.3, 1.8, and 1.04 m, with a mere weight of 170 kg, driving speed of 100 km/h, and maximum speed of 140 km/h.

(8) **Solar energy technology application in the Expo Site**

The communal facilities at the Expo site such as street lights, lawn lights, park lighting, and mosquito killing devices made good use of solar technology, shown in Figs. 2.13, 2.14, and 2.15.

Fig. 2.13 Solar energy black-light mosquito-killing device

Fig. 2.14 Solar decorative lights

Fig. 2.15 Solar intelligent recycling bins outside the London's UBPA case

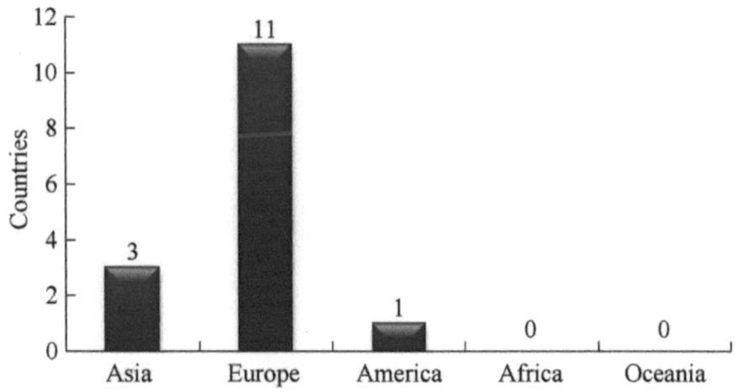

Fig. 2.16 Applications of solar energy in the Expo (Statistics by continents)

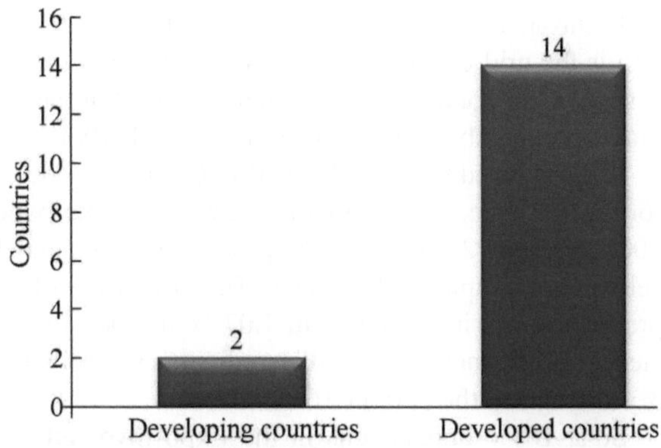

Fig. 2.17 Applications of solar energy in the Expo (Statistics by developed and developing countries)

2.1.1.4 Statistics of Solar Energy Applications in the Expo

The applications of solar energy in the Expo are shown in Figs. 2.16 and 2.17.

The application of solar energy technology in the G8 and G20 countries (as shown in Figs. 2.18 and 2.19), accounts for 60 and 85 % of the global GDP respectively. Being global economic leaders, G8 and G20 also lead the world in development of new energies.

2.1.1.5 Status of Solar Energy at Home and Abroad

At present, 136 countries around the world are promoting solar power generation technology, of which 95 countries are researching and developing large-scale solar power generating equipment and solar applications. In 2005, the total output of the world's solar cells production reached 2,000 MW. Japan takes the lead in terms of solar energy development, whose solar cells output account for about 50 % of the world's total production. Global solar PV capacity in 2009 increased by 44 % and new installed capacity reached 6.4 billion MW, making the global installed capacity to more than 20 billion MW.

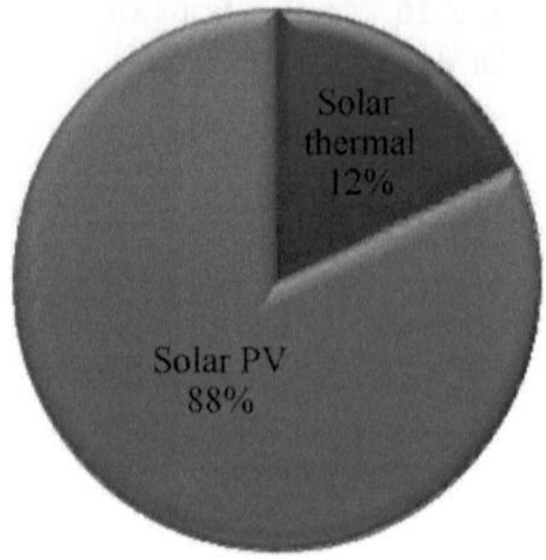

Fig. 2.18 Applications of solar energy in G8 national pavilions

Eighty-five percent of the world's total installed capacity of PV power generation is the grid-connected PV power generation system, 90 % of which is the roof system. Grid-connected PV systems in the European countries, Japan, the United States are mainly building integrated PV (BIPV).

In 2007, production of PV cell in China exceeded (that of) Germany and Japan for the first time, ranking first in the world. The production continued to increase in 2008, reaching 2 GW. Over the past 5 years the annual growth rate of output of PV cells in our country is 1–3 times. The proportion of solar PV cells output in global production has increased from 1.07 % in 2002 to nearly 15 % in 2008. The efficiency of commercial crystalline silicon solar cells has increased from 13 to 14 % 3 years ago to the current 16–17 %.

Solar BIPV applications in the Expo involved a number of key technologies, many of them China's own innovative technologies, such as the combination of solar energy with building technology, BIPV component technology, brick-type PV module, large area Transmission-type PV module, PV waterproof module, shaped PV modules, high-power inverters and efficient combination technology, large-scale PV building control technology, and systems engineering technology. Currently, China has introduced a number of supportive policies on the promotion of BIPV projects and applications. BIPV of large public buildings is the international trend and will be the biggest growth point of the future domestic market as well.

2.1.2 Wind Energy

2.1.2.1 Overview

Wind energy is the kinetic energy generated by a large amount of air movement on the Earth's surface. It is a converted form of solar energy. Wind energy resource is determined by the density of wind energy and the annual accumulative hours of available wind energy. It is estimated that the total amount of wind energy worldwide is approximately 130 GW, 10 times more than the total amount of global exploitable water energy. Wind has long been utilized by man, mainly for pumping water and milling grains through windmills. But at present, people are interested in

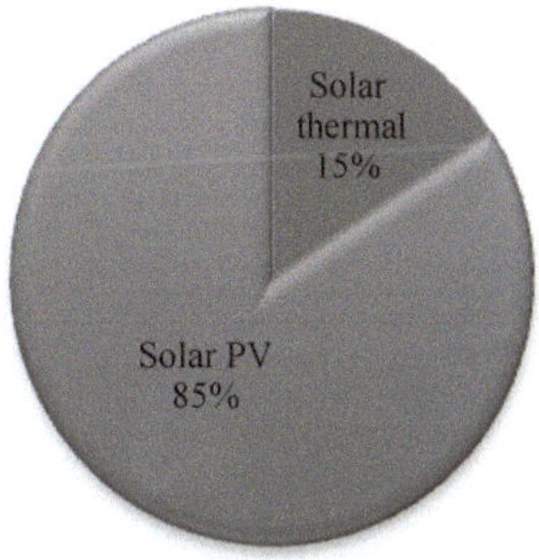

Fig. 2.19 Applications of solar energy in G20 national pavilions

Fig. 2.20 Wind power generation

how to make use of wind power to generate electricity. Figure 2.20 shows wind power generation in Inner Mongolia Huitengliang.

(1) **The Principle of Wind Power Generation**

The principle of wind power generation is to drive the windmill blades to rotate by wind and then to increase the speed of rotation through the speedup gearbox, to prompt the generation of electricity. According to the existing windmill technology, when wind speed reaches approximately 3 m/s (breeze speed), it begins to generate power. The advantages of wind energy explain why it is the fastest-growing energy source in the world. By using clean fuel source, wind power generation does not pollute the air nor generates radiation. Currently, wind power generation is becoming popular throughout the world.

Wind turbines include large fans and small fans. Wind turbine is composed of a handpiece, rotor, empennage, and blades. Each part functions as follows: blades receive the wind power and convert it into electric energy; empennage keeps the blades facing the direction of wind all the time so as to obtain the maximum wind

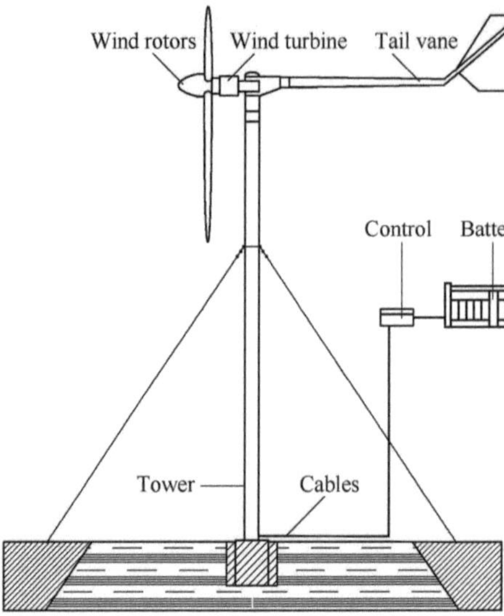

Fig. 2.21 The working principle of wind power generation

energy; rotor makes the handpiece rotate slickly to achieve the direction adjustment by the empennage; the rotor of the handpiece is a permanent magnet and stator winding carves up the line of magnetic force to generate electric energy. The principle of wind power generation is illustrated in Fig. 2.21.

(2) **Classification of Wind Turbines**

Although there are a variety of wind turbines, there are generally two kinds: (1) horizontal axis wind turbine (HAWT) (as shown in Fig. 2.22), in which the rotor shaft of the turbine blades are parallel to the wind direction; (2) vertical axis wind turbine (VAWT) (Fig. 2.23) in which the rotor shaft of the turbine blades are perpendicular to the ground or to the direction of airflow.

Horizontal axis wind turbine (HAWT)

HAWT can be again classified into the type of lifting force and dragging force. Lifting force wind turbine rotates more quickly than dragging force wind turbine. In wind power generation, HAWT with lifting force is widely used. Most HAWTs are equipped with wind-facing devices that will rotate to the change in wind direction. For small wind turbines, tail vane is used in this device, while for large-scale wind turbine, transmission device consisting of wind direction sensors and actuating motor are used.

HAWT is divided into upwind turbine with the wind rotor in front of the tower and downwind turbine with the wind rotor behind the tower. There are many types of HAWT. Some are installed with wind rotor of blades reversal and some with several wind rotors installed on one tower so as to reduce the cost of the tower at certain output power; still for some type of turbines, swirls are produced around the wind wheel to centralize air flow and therefore increase air velocity.

Fig. 2.22 Horizontal axis wind turbine (HAWT)

Fig. 2.23 Vertical axis wind turbine (VAWT)

Vertical axis wind turbine (VAWT)

Compared with the HAWT, VAWT does not have to face the wind when the direction of wind changes which is an advantage. It not only simplifies the structure, but also reduces the resistance of wind rotor when facing the wind.

There are several types of VAWT rotating utilizing the resistant power for rotation, including the one with the wind rotor made of plate and rotor which is a pure resistance device; the S-shaped windmill has a part of the lifting force but is still a kind of resistance device. These devices have bigger starting torque, but the tip speed ratio is low. However, at a constant wheel rotor size, weight, and cost, such devices have lower output power.

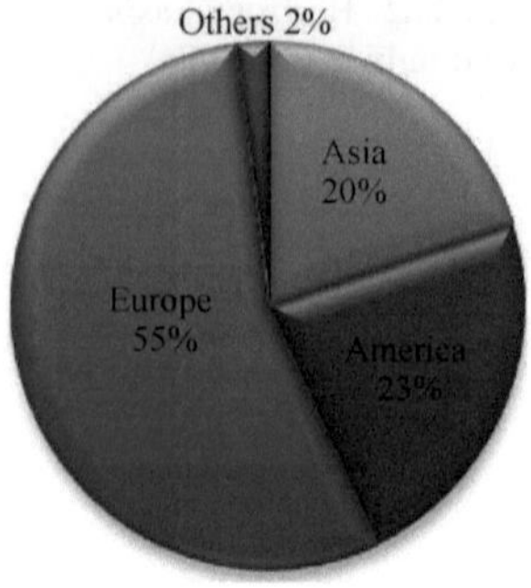

Fig. 2.24 Distribution of global installed wind power capacity (2008)

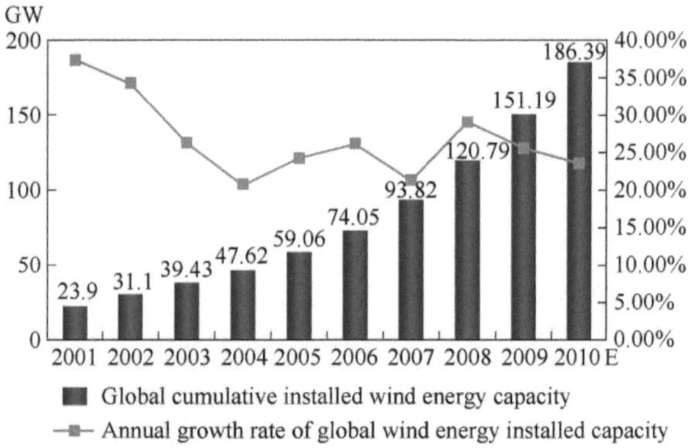

Fig. 2.25 Growth of global cumulative installed capacity (2001–2010)

(3) The Status of Wind Power at Home and Abroad

Over the past decade, wind power generation stations have grown rapidly. Since 2001, the capacity of the installed wind power worldwide has increased at the rate of 20–30 % per year. In 2008, the newly increased installed wind power capacity worldwide was 27,050 MW, making the total global installed capacity to 120,000 MW, up by 28.8 % compared to 2007. Distribution of global installed wind power capacity in 2008 is shown in Fig. 2.24. Global growth of wind power installed capacity from 2001 to 2010 is shown in Fig. 2.25.

Wind power in China started in 2003, and has grown rapidly especially after the launch of the Wind Power Concession Project. In 2006, a total of 1,454 wind turbines were produced in China, with the newly added installed capacity reaching 1,337 MW, not including Taiwan, exceeding the total cumulative capacity of the past 20 years, ranking only after the US, Germany, Spain, and India. In 2008, China doubled its installed wind power capacity by adding 6,300 MW of wind turbines, ranking second globally in terms of the newly increased capacity, second only to the United States. By the end of 2009, the total installed capacity in China was 21,290 MW, with a year-on-year growth of 74.36 %, surpassing Spain, ranking third in the world and among the top wind power countries with the installed wind power capacity over 10,000 MW. Figure 2.26 reflects the installed wind power growth in China since 2001.

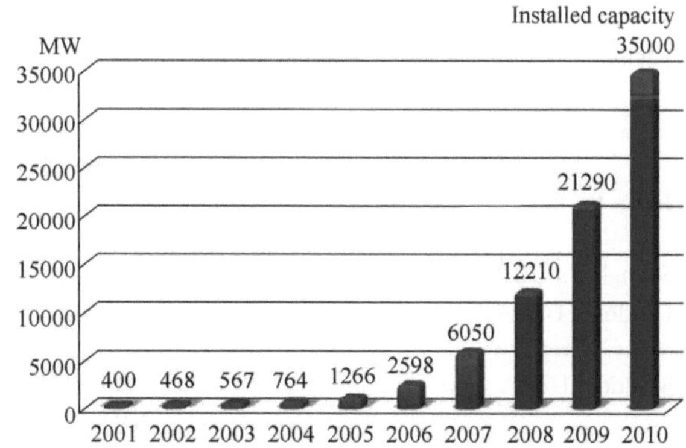

Fig. 2.26 Installed capacity in China (2001–2010) (MW). *Source* www.wwindea.org

The above data were collected before Dec. 31, 2009, when China's wind turbine grid connection rate reached 76 %. According to the types of China's wind power installed turbines in 2008, the capacity of double-fed turbine accounted for a huge proportion. Owing to a shortage of components, direct drive wind turbine accounts for only a small proportion. With the impact of supply of the above components, in the next few years, direct drive and double-fed wind turbines will coexist.

2.1.2.2 Wind Power Application in the Expo

Wind power applications in Expo 2010 Shanghai can be divided into three categories: First, the practical application of wind turbines, second, wind turbine components, and third, display of wind power enterprises and wind power projects. Detailed information is given in Table 2.2. New energy utilization by classification is shown in Fig. 2.27.

(1) **Wind Power Application in the Expo** The Turbines applied in Expo 2010 Shanghai include VAWT and HAWT. According to the statistics, in terms of the types of turbines applied in the Expo, VAWTs accounted for 67 %, while horizontal types accounted for 33 %.

Demonstration of wind power enterprises: domestic enterprises focus on technology-based demonstration, while foreign enterprises focus on the complete turbines demonstration, which are from large wind turbine manufacturers.

The survey showed that the pavilions with wind power applications included the India pavilion, Japan pavilion, Shanghai UBPA case pavilion, London UBPA case pavilion, Madrid UBPA case pavilion, Pavilion of Future, the Expo Ferry, Mobile base stations, and the State grid Pavilion. Many wind turbines used in the above pavilions were small VAWTs, which is due to the fact that chaotic wind often exists in the city. VAWTs need not face against the wind when the wind direction changes, which not only features the simplified design structure, but also, the relatively small gyroscopic force when the wind wheel faces the wind results in

Table 2.2 Wind turbines display and applications in Expo 2010

Pavilions	Category	Wind turbine type	Displayed items and features
India Pavilion	Practical use	Vertical axis wind turbine	Lift-type wind turbine, H type, 3 kW
Japan Pavilion			Resisting type wind turbine wind-solar hybrid
Shanghai's UBPA Case			Lift-type wind turbine, H
London's UBPA Case			Lift-type wind turbine, H-type, wind-solar hybrid
Madrid's UBPA Case		Horizontal axis wind turbine	5 turbo generators
Pavilion of Future		Vertical axis wind turbine	Lift-type wind turbine, H type
Expo Ferry		Horizontal axis wind turbine	Inland riverboat wind turbine
Expo Mobile Base		Vertical axis wind turbine	Lift-type wind turbine, H type, Peak power output 3,000 W/h
State Grid Pavilion		Horizontal axis wind turbine	Breeze started vertical axis wind turbine
Sweden Pavilion	Wind turbine components		Super bearing
China Pavilion			Pitch and Yaw bearings
Joint Pavilion of International organization			Bamboo composite wind blades
Pavilion of Urban Planet			Gearbox
Pavilion of future	Wind power companies display		Wind power control device
Theme Pavilion			Large wind turbine equipment
Portugal Pavilion			Wind farm management
Denmark Pavilion			The world's first wind power company
Netherlands Pavilion			Business display, the National Wind Resource Development
Meteoworld Pavilion			Business display
Shanghai Donghai Bridge 100 MW offshore wind farm	Provide green energy for world Expo		The total installed capacity of 102,000 kW

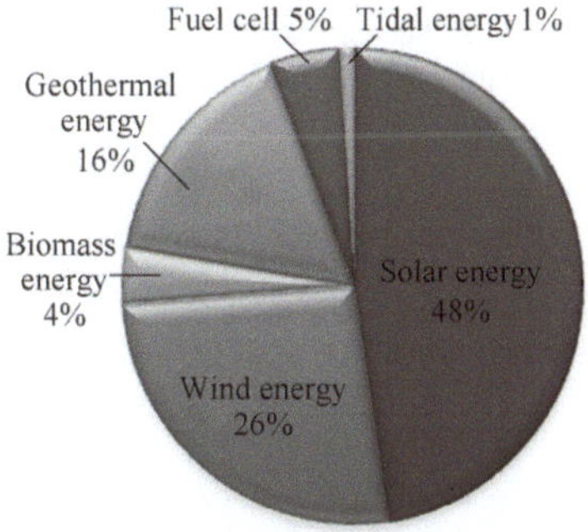

Fig. 2.27 New energy utilization by classification

Fig. 2.28 Wind-solar hybrid power supply system in London's UBPA case

high safety in running. The above features allow the vertical wind turbine to have superiority in application in urban buildings than the horizontal axis. New VAWTs with low wind speed, or even with breeze starting, were eye-catching in the Expo.

Wind-solar hybrid power supply system in London's UBPA case is shown in Fig. 2.28.

The wind-solar hybrid power supply garden lamps from Panasonic displayed outside the Japan pavilion (Fig. 2.29) was a typical resistance wind wheel (Savonius), which have the following advantages: (1) Adapt to the frequent change in wind direction; (2) easy starting, with low wind speed; (3) safety in design; (4) marvelous appearance, with the effect of landscaping, different in design from the ordinary type; (5) convenient height, easy for erection and maintenance; (6) zero noise; (7) match with wind status at the roof of ordinary building height. The disadvantage is its relatively low material utilization; for the given structural material, the front surface of the wheel is small.

Fig. 2.29 Scenery wind and solar lamp outside Japan Pavilion

Fig. 2.30 Vertical axis wind turbines on the top of the India Pavilion

The typical application of building integrated wind power was at the India Pavilion. It used a VAWT (Fig. 2.30), with rated power of 3 kW, generating AC of 3 kW at 12 m/s of wind speed. The wind power generation system complemented the solar power systems on the rooftop of its pavilion. Wind and solar hybrid power supply system was also used in the Expo mobile base station (Fig. 2.31). The fans used in the station had peak power output of 3,000 W/h; the usual generating capacity and amount of installed solar power exceeded the total electricity consumption in the base station by over 1/3, which ensured the self-sufficiency in daily light supply within the base station.

Small HAWTs were used in the State Grid Pavilion, Expo Ferry, and Madrid's UBPA case pavilion (Fig. 2.32). Five wind turbine generators were installed in the Air tree system in Madrid's UBPA case pavilion (Figs. 2.33 and 2.34). With the Air

Fig. 2.31 Wind-solar hybrid power supply system in the mobile base of the Expo

Fig. 2.32 300 W Horizontal axis wind turbine on the Expo Ferry

tree system, the total generating capacity reached 1.7 kWh. Wind turbines can generate electricity even in at wind speed of less than 2 m/s. All the power generated is fed into the Shanghai power grid.

(2) Wind Turbines Components

Wind turbine is composed of rotor, gearbox, generator, yaw device, control system, and tower. However, the most critical component of wind turbines is the blades. Blade design is especially important because proper design of the blades will ensure the generating unit to obtain the desired power. Besides, fatigue characteristics of blades are also critical, which is related to the materials of the blades. The Joint Pavilion of international organization demonstrated bamboo composite wind blades (Fig. 2.35), which showed better performance at 30 % higher than that of the widely used glass fiber reinforced plastic; it also reduces energy consumption in production and can be recycled. Aside from meeting the strength requirements,

Fig. 2.33 "Air trees" in Madrid Pavilion

Fig. 2.34 Wind turbine topped the "Air tree"

the blade is lighter in weight, longer than the old type, and has higher output power with lower comprehensive cost. This kind of blade has been put to use in Zhangbei Power Plant in northern China and has begun to be connected to the grid.

The bearings of the turbine were displayed in the Sweden and China Pavilions. Pitch bearings and yaw bearings were shown in China Pavilion (Fig. 2.36). Pitch bearings can adjust blade angle, keep the speed constant, and the generated power stable; yaw bearings track wind direction, align the windward side, and improve power generation efficiency. The 1.5 MW wind yaw bearings and pitch bearings manufactured by Ma'anshan Fangyuan Slewing Ring Limited liability Company have superior performance. SKF super bearings displayed in the Sweden Pavilion (Fig. 2.37) can be used for large-scale wind turbines and are characterized by:

Fig. 2.35 Wind blades made of bamboo composite

Fig. 2.36 Pitch and yaw bearings showed in China Pavilion

(1) Lightweight, small friction allowing wind turbines to run by breeze;
(2) The bearings optimize the performance of the units and maximize their life span;
(3) The novel bearings receive a special treatment of corrosion-resistant surface, and special oils and fats used to limit the indentation and damage of the surface;
(4) The bearings are equipped with innovative automatic lubrication system, which can allocate the appropriate amount of oil to all the lubrication points during a 6-month period, thereby reducing maintenance cost and oil consumption;
(5). Patent sealing solutions are used so as to avoid oil leaks and water intrusion, and can cope with the deformation bearing ring result from overload and effect of internal fat overpressure of the bearing.

Fig. 2.37 The Super bearing displayed in Sweden Pavilion

Fig. 2.38 Wind turbine pitch —and yaw-control system

In addition, by adoption of an innovative double-layer surface coating design, the bearings can also provide lasting corrosion protection even in bad weather.

Gearbox as part of wind turbine is also displayed in City Earth Pavilion, where Bosch showcased the gear box it manufactured (Fig. 2.38). Quiet and reliable REDULUSGPV-type generator gear box (Fig. 2.39) and compact REDULUSGPV-D-type differential gearbox became the company's flagship product used in wind turbines. The gearbox has high strength anti-wear bearings, lubrication performance, compact gear combination, big gear ratio range, low noise, accurate and timely monitoring, etc.

(3) **Wind power companies in the Expo**

In London Case Pavilion, ZK Energy Technology Co., Ltd displayed not only vertical wind turbine products, but also a full permanent magnetic suspension wind and solar power generation technology (see Fig. 2.40). Full-permanent magnetic

Fig. 2.39 REDULUSGP type gear box generator

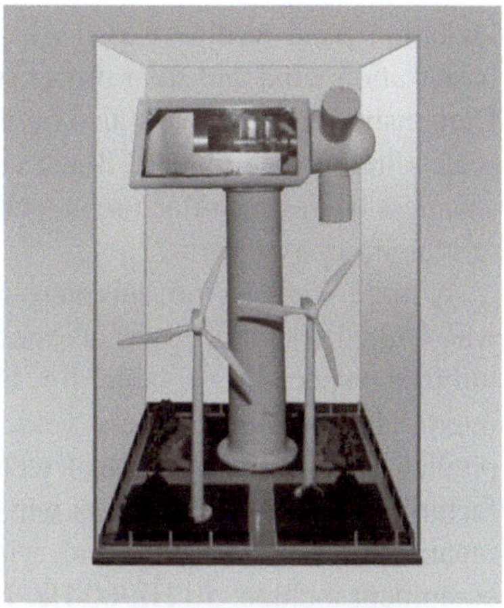

Fig. 2.40 Permanent magnetic suspension wind power turbine in London Case Pavilion

Fig. 2.41 Wind-solar hybrid power supply system in communication base station

Fig. 2.42 Direct drive wind turbine of EWT Company

levitation wind turbine was the highlight of the display, while a communications base station wind and solar power supply system (see Fig. 2.41) in The Low-carbon Demonstration Town was also on display. The Theme Pavilion showcased a large-scale wind turbine Integral Blade with Siemens technology, which can achieve full, seamless blades manufacture in a close process. The blades were made of glass fiber reinforced epoxy resin.

A large wind power company in China displayed in the Meteoworld Pavilion was United Electric Power Technology Co., Ltd. The company developed wind turbines especially suitable for abnormal weather conditions such as marine, intertidal, high altitude, low velocity, cold, sand; it also developed wind power prediction and remote control technology and conducted R&D on green manufacturing technologies such as wind turbine weight reduction, the key components repair and life extension, and wind turbine blade material recycling. Its R&D equipment include: 50 HZ/60 HZ 1.5 MW constant frequency variables pitch wind turbine; high-altitude low-wind speed model leaves 1.5 MW; 3 MW offshore wind turbines; Electro-hydraulic/solenoid coupled synchronization wind turbines.

Netherland Pavilion showed the utilization of wind energy resources at home (Fig. 2.42). EWT's direct drive technology, DW90-2.0 MW, DW54-750/900 kW large wind turbines were the most attractive highlight.

Portugal Pavilion introduced EDP Company (Fig. 2.43), the world's second largest wind power company located in its country. The company is very professional in wind farm operation and management with operations throughout Europe, the United States, and Brazil. Its installed capacity is 5.8 GW. In 2010 the company achieved its annual target of 10.8 GW. EDP Company worked with the Principle Power Company and A. Silva Matos Company in developing an offshore wind farm project—The Sea Floating. The wind power generating platform with innovative design can be installed on water surface with depth of more than 50 m, allowing exploration of renewable energy that could not be reached previously.

葡萄牙EDP-Renováveis公司负责管理遍及全球的业务网络，在欧洲、美国和巴西均有风电场，其装机容量为5.8吉瓦，预计2010年实现10.8吉瓦的目标。这些风电场由位于葡萄牙波尔图市的调度中心统一控制，该中心管理风力涡轮机全球第二大业务网络。

A empresa portuguesa EDP - Renováveis é responsável pela gestão de uma rede mundial de parques eólicos na Europa, Estados Unidos da América e Brasil, com uma capacidade instalada de 5,8 GW, prevendo-se para 2012 uma meta de 10,8 GW.

Fig. 2.43 EDP Company in Portugal Pavilion

Fig. 2.44 Donghai bridge offshore wind farm

(4) Expo offshore wind farm

China's first large offshore wind farm—Shanghai Donghai Bridge Offshore Wind Farm (Fig. 2.44)—is one of the offshore wind power demonstration projects determined by National Development and Reform Commission. Shanghai Donghai Bridge Offshore Wind Farm is made of 34 sets of 3,000 kW turbines from Sinovel Wind Power Technology Co., Ltd., with a total installed capacity of 102,000 kWh. Its annual generation capacity achieved 267,000 MW, providing green power from the sea to the World Expo.

2.1.2.3 Analysis on Wind Power Application in the Expo

The study on Expo 2010 Shanghai showed that cases of wind power application are less than that of solar power application and the use of small urban-type wind turbine is relatively fewer. Hence the development of urban-type wind turbines still

Fig. 2.45 Wind power applications in the Expo (Statistics by continent)

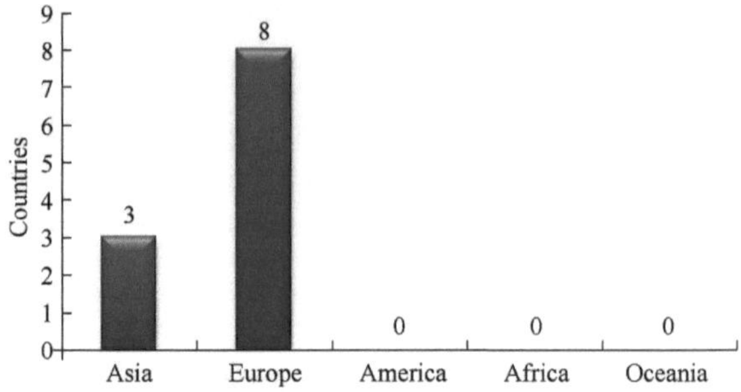

Fig. 2.46 Wind power applications in the Expo (by developed and developing countries)

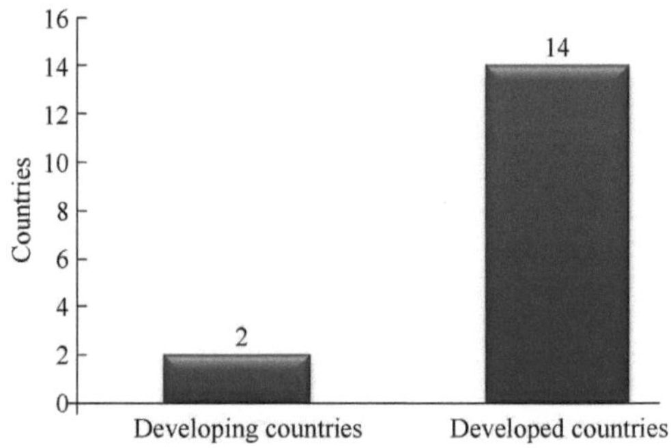

has space for improvement. The urban-type has good prospect in application, and is beneficial to alleviating the tension of urban energy.

The demonstration of wind turbine products and technologies were mainly seen in the pavilions of the Asian and European countries. Other pavilions did not display wind turbines due to various reasons. Figures 2.45 and 2.46 show wind power applications in the Expo.

Donghai Bridge Offshore wind farm, with installed capacity of 100 MW, is expected to generate electricity of 267 million kWh and meet the demand for electricity of more than 20 million residents of Shanghai a year. After completion, Shanghai can save 100,000 tons of coal annually, and reduce 200,000 tons of carbon dioxide emission, 1,200 tons of sulfur dioxide, and about 670 tons of nitrogen dioxide.

2.1.2.4 Recommendations on Wind Power Promotion

In the pavilions of the Expo site, cases of large wind turbines utilization were shown through models and texts. Large wind turbines have developed rapidly both at home and abroad, expanding from land to ocean, from large type to huge type. Most of the small wind turbines used in the Expo site were the vertical axis type wind turbines which require high startup speed, of which China is accelerating its

R&D and gradually keeping in line with US and European countries, while the demand from abroad is converted into that of landscape and environment-friendly type. Our recommendation is to speed up the R&D in light breeze-start products and its industrialization pace to match with the building greening, urban landscaping and lightening.

2.1.3 Biomass Energy

(1) **Introduction to Biomass Energy**

Biomass refers to all kinds of organisms formed through the process of photosynthesis, including all animals, plants, and microbes. The so-called biomass energy is the captured solar energy stored in the biomass in the form of chemical energy. In other words, it is a kind of energy with biomass as the carrier. It comes directly or indirectly from the photosynthesis of green plants and converted into solid, liquid, and gas conventional fuels that can never be used up. It is one of the renewable energies and the only renewable carbon source as well. The primary energy source in biomass energy is derived from the Sun. In a broad sense, biomass energy is a manifestation of solar energy. Biomass converted from the solar energy is stored in the living organisms such as plants, animals, and microbes. All the energy sources except mineral fuels of organisms, including wood, forest wastes, agricultural wastes, water plants, oil plants, urban, and industrial garbage and animal wastes, etc., belong to the biomass energy. Figure 2.47 shows biomass energy utilization. At present, many countries are actively developing the biomass energy to put it into use.

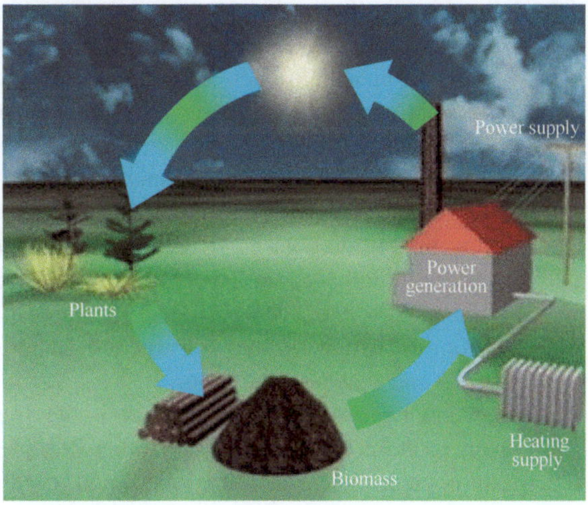

Fig. 2.47 Biomass energy utilization

Biomass resources are abundant in China, but it is not evenly distributed in different provinces. More than half of the biomass resources are concentrated in the nine provinces: Sichuan, Henan, Shandong, Anhui, Hebei, Jiangsu, Hunan, Hubei, and Zhejiang, while a relatively small amount is in the vast northwest regions and other provinces. Annual straw production capacity is 800 million tons, most of which is burned or buried except for a small amount used as household firewood and feed. Fuel wood output is 250 million tons. Besides, there are plenty of wastes from man, and livestock organic waste residues from industrial emission.

(1) **Classification of biomass** Based on different sources, the biomass that is suitable for energy utilization can be divided into: forest resources, agricultural resources, domestic sewage and industrial organic wastewater, urban solid wastes, and livestock manure.

(i) *Forest resources:* Forest biomass resources refer to the biomass energy resources provided during the processes of forest growth and forestry production. They include firewood forest, scattered wood and residual branches, leaves and bits of wood during tending of woods and intermediate cutting, twigs, sawdust, top wood, slab produced during the timber harvesting, and process as well as garbage of the forestry byproducts such as shuck and kernel, etc.

(ii) *Agricultural resources*: Agricultural biomass resources refer to the agricultural crops (including energy crops), wastes produced during the process of agricultural production like straws residual after crops harvesting (corn straw, sorghum stalks, wheat straw, rice straw, soybean, cotton stalk, etc.), agricultural processing wastes like remainder of rice husk produced in the process of agricultural production. Energy plants generally refer to various kinds of plants that are used to provide energy, usually including herbal energy crops, oil-bearing crops, plants for making hydrocarbons and aquatic plants.

(iii) *Domestic sewage and industrial organic wastewater*: Domestic sewage is drained water from urban residents' wastewater, commerce and service's wastewater, such as cooling water, bath drainage, washing drainage, clothes washing drainage, kitchen drainage, fecal sewage, etc. Industrial organic wastewater mainly contains sewages drained from the production process of alcohol, wine-brewing, food, medicine, papermaking, and slaughtering industries, which are rich in organic matters.

(iv) *Urban solid wastes*: Urban solid wastes are mainly composed of solid wastes such as urban living garbage, commercial, service industry garbage and some construction garbage, etc. Its composition is complicated, which is influenced by the factors like average living standard of local residents, energy consumption structure, urban construction, natural conditions, and traditional customs and seasonal variation.

(v) *Livestock manure*: Livestock manure is the generic term for livestock feces, which is the converted forms of other biomass (mainly grain, crops straw, pasture, etc.), including animal feces and urine and its mixture with bedding.

(2) **Characteristics of Biomass Energy**

(i) *Reproducibility*: Biomass energy, like wind energy and solar energy, is a renewable resource, which can be reused through the photosynthesis of plants and is abundant enough to ensure the everlasting use of the energy.

(ii) *Less Pollution*: Biomass contains low sulfur and nitrogen, so it generates little SO_X and NO_X (nitrogen oxides) during combustion process. When Biomass is used as a fuel, the amount of carbon dioxide it requires during its growth is equivalent to the amount of carbon dioxide it lets out, thus the net carbon dioxide emission to the atmosphere is kin to zero, which can effectively reduce the greenhouse effect.

(iii) *Wide Distribution*: In regions with a shortage of coal, biomass energy can be widely used.

(iv) *Huge Storage*: According to biologists estimate, there are about 100–125 billion tons of terrestrial biomass and 50 billion tons of ocean biomass per annum. The annual production of biomass is far more than the world's demand for all energies, equivalent to 10 times of the current total global energy consumption. In 2010 the biomass resources that could be developed in China amounted to 300 million tons.

(3) **Biomass energy utilization:** There are three ways in using biomass energy: direct combustion, thermo-chemical conversion and biochemical conversion. Direct combustion of biomass will be the dominant way of biomass energy utilization for a long time in the future in China. The program of making transformation of the traditional firewood stove with the thermal efficiency of only about 10 % into the energy-saving stove with the promoting efficiency up to 20–30 %, is listed as one of the key tasks in the new energy construction in rural areas. Thermochemical conversion of biomass is the technology to make biomass gasification, carbonization, pyrolysis, and catalytic liquefaction at a certain temperature and condition, so as to produce gaseous fuels, liquid fuels, and chemical substance. Biochemical conversion of biomass includes biomass-methane conversion and biomass-ethanol conversion. Methane conversion is the technology to put organic matter in the anaerobic environment, to produce methane through microbial fermentation as the main component of a combustible mixture of gases; Ethanol conversion is to produce ethanol through fermentation by using carbohydrate, starch, and cellulose as raw materials.

Biomass energy has always been an important energy source for human's life, which is only second to coal, oil, and gas and ranks fourth with regard to total energy consumption. It plays an important role in the whole energy system. Experts estimate that biomass is likely to become an integral part of sustainable energy system in the future. By the middle of the next century, a variety of biomass alternative fuels produced with new technology will account for more than 40 % of global energy consumption.

Fig. 2.48 Microalgae oil plant of ENN group the in China Pavilion

Fig. 2.49 Microalgae railings shown in the Shanghai Case Pavilion

2.1.3.1 Biomass Energy Utilization in the Expo

In the north corner of the London ZED Pavilion, there is a biomass boiler which has been placed in a separate room. When the tourists finished eating, biomass boilers can immediately degrade the leftovers into electricity and heat. It can dispose the restaurant's one-off plates, forks, and a variety of food wastes by biological anaerobic degradation. When the degradation is completed, the final "products" can be used as fertilizers for the green vegetation on the northern slope of the roof.

While most companies are still investing a lot of fund and human resources on how to reduce carbon dioxide emissions or sequestrate carbon dioxide, a private enterprise named China ENN Group displays a new technology—"microalgae oil"—to the visitors in China Pavilion, and also exhibits "microalgae railing" in the Shanghai Case Pavilion (Figs. 2.48 and 2.49). It can absorb carbon dioxide with the cultivated microalgae, and ultimately produce bio-diesel, carotenoid, and other derivatives through photosynthesis. It can change carbon dioxide into what can be utilized.

The principle of the microalgae-oil is through photosynthesis of the microalgae, to change the carbon dioxide produced in the chemical production process into biomass of microalgae so as to fix the carbon elements, and then through the self-induced reactions to change the substance of carbon into oil. After that the oil inside the algae cells will be transferred outside the cells by physical or chemical methods, and lastly, with further refining process, bio-diesel is produced.

According to reports, compared with corn, soybeans, and rapeseed, algae cultivation requires a relatively small area with short growth cycle. It takes just 2 weeks from birth to oil production, while oil crops generally take several months for the same process. Besides, oil production of microalgae per unit area is several hundred times that of corn; which is 15,000–80,000 L of bio-diesel per ha.

2.1.3.2 Economic Returns of Biomass Energy

About 70 % of the population in China lives in rural areas, where straw and firewood and other biomass fuels are the major fuels for daily life. Although coals and other commercial fuels used in rural areas are increasing rapidly, biomass energy still plays an important role. In 1998, the total energy consumed in rural areas amounted to 365 million tons of coal equivalence, of which straw and firewood represented 207 million tons of coal equivalence, accounting for 56.7 % of the total. Therefore, promoting biomass energy technologies to provide energy for the daily use and production in rural areas is the best way to eliminate poverty in these areas and help to achieve the target of building a well-to-do society.

China boasts its abundant mountainous resources and therefore there is great potential in the development of forestry biomass energy: about 44.04 million ha of barren hills, wasteland can be used for cultivation of energy forests; nearly 100 million ha of desert land, mines, oil fields, and other lands can be used as energy forests; about 6 million ha of woodlands and 53.12 million ha of low-yield forests can significantly increase the amount of forest resources. According to *National Forest Biomass Energy Development Plan* being compiled, by 2020 up to 20 million ha of biomass forests will be achieved, the annual transformed forestry biomass energy can replace the petrochemical energies of 20.25 million tons of coal equivalence, accounting for 3 % of all the renewable energies.

2.1.4 Fuel Cells

2.1.4.1 An Overview of Fuel Cells

Fuel cell is a kind of chemical cell that needs to be replenished constantly. Viewed from its structure and working principle, fuel cells are no different from the conventional chemical cells; Substantially, fuel cells are not conventional chemical cells which are energy storage device, they are, however, kind of power generation device. The fuel cell converts the chemical energy from the fuels and oxidizers directly into electricity. It is a newly developed fuel-cell power generation technology after the hydroelectric power, thermal power, and nuclear power generation technologies, which is characterized by its high efficiency and environmental friendly advantage. Compared with the piston engine, the fuel cell has a longer history but developed slowly in the past years.

In 1800, Volta invented the first battery.

In 1802, Davy trial-made the carbon-oxygen cells with carbon and oxygen as fuels and nitric acid as electrolytes. He pointed out that it is possible to make fuel cells.

In 1839, Grovethe electrolyzed water to have reverse reaction. He used dilute sulphuric acid to make hydrogen react with oxygen, which produced the current, and by doing the experiment he discovered the working principle of hydrogen-oxygen fuel cell.

In 1894, Ostwald made an analysis and pointed out that the generating efficiency of fuel cells could reach 50–80 %, while the generating efficiency of the heat energy was less than 50 % due to the limitation of Carnot Cycle. But for a long period of time after that, limited by the applied materials and design, the fuel cell demonstrated such low dynamic performance that the experimental research developed rather slowly with little achievements.

From the 1950s, came a turning point in the development of fuel cells. Bacon of the Cambridge University conducted a long-term study on oxyhydrogen alkaline fuel cells (AFC) which was very productive. He successfully made a 5 kW AFC that can last 1,000 h. This was the first practical fuel cell. The achievements that Bacon made successfully established the technical idea for the modern fuel cells. In the 1960s, the race in space technology between United States and the Soviet Union needed cells with high-performance badly to provide power in the space. By analysis and assessment of different batteries, NATO chose fuel cells, by adopting Bacon's technology they developed the fuel cell used for the Apollo space craft. The great success of the fuel cell used in space flight spurred the development boom of the fuel cells. In the 1970s, the energy crisis was arisen after the Middle East war and people paid much attention to fuel-cell power generation technology. The United States, Japan and some other countries all made long-term plans in developing fuel cells. Some developed countries led by the US embraced the development of fuel cell power stations for civil use and established a number of tentative medium and small power stations. At the same time, the United States and Japan started to transfer the surface station research to high temperature fuel cells with better properties. After the 1990s, fuel cells were further developed. The focus of the research turned to the solid fuel cells, namely proton exchange membrane fuel cell or PEMFC. Because it gives off low emission or zero emission, the fuel cell is becoming an increasingly hot spot in vehicle industry.

Fuel cells can either be used as auxiliary power unit (APU) for vehicles or as the engine to supply power to the motor-driven vehicles. When it provides APU for a car, the car still uses internal-combustion engine where some fuel is directly

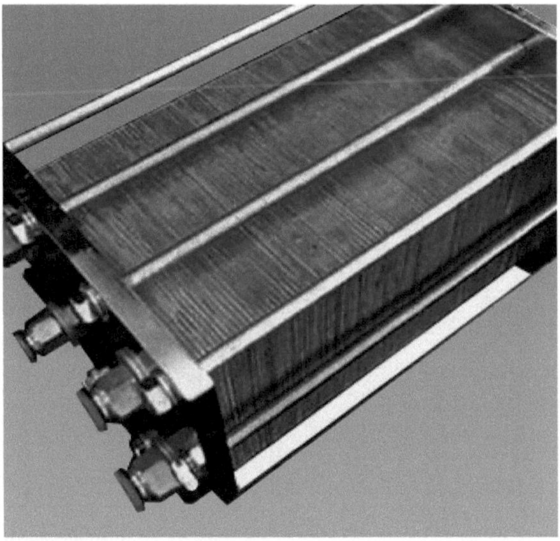

Fig. 2.50 Fuel cells

transferred to electricity through fuel cells to provide for the vehicles. This is more efficient than the pattern of traditional engine plus battery pack. The fuel cells-driven cars mentioned above usually refer to the cars that use fuel cells as its main motivation. In that case, fuel cells can be coupled with storage battery which is used only at the automobile starting stage requiring peak power. In addition, storage battery can also be used to absorb the braking power of the cars as well. Figure 2.50 shows fuel cells.

Based on different electrolytes, fuel cells can be divided into AFC, phosphoric acid, the fuel cell (PAFC), molten carbonate salt fuel cells (MCFC), solid oxide fuel cells (SOFC) and PEMFC, etc. Not confined by Carnot cycle, fuel cells have such advantages as high energy conversion efficiency, cleanness, pollution-free, and low noise that they can be made into the modules structures, with power density higher than that of conventional batteries, which is suitable for both centralized and distributed power supply. The main application in cars is the PEMFCs using hydrogen as its fuel.

Figure 2.51 shows the diagram of the PEMFC structure. At present, the prominent feature of fuel cells in structure is they can only choose pure hydrogen as fuel and platinum (noble metal) as catalyzer for electrode reaction. The working temperature should be between 20 and 80 °C.

The reaction on the Anode side of the cell is formulated as: $H_2 \rightarrow 2H^+ + 2e^-$, on the cathode side of the cell is $O_2 + 4H^+ + 4e^- \rightarrow 2H_2O$. The hydrogen fuel cells vehicles have the advantage of zero-emission or low emission. In addition, compared with the traditional piston engine, as fuel cells have no moving parts, they will not produce any vibration or noise, besides, the fuel is more efficient. Compared with battery pack in the traditional electric car, fuel cells get longer driving mileage, lighter weight, longer service life, and quicker refuel time.

Fig. 2.51 Working principle of the proton exchange membrane fuel cell

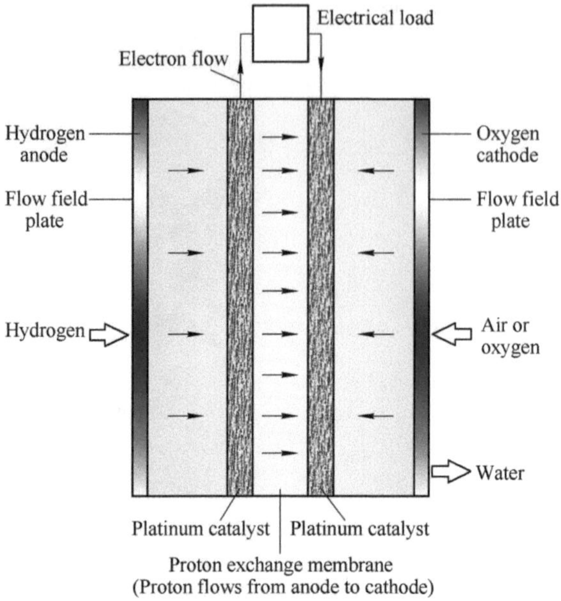

Since the beginning of the twenty-first century, energy crisis and environmental pollution have become a challenge facing all countries. Fuel cell is a power generation device which converts chemical energy directly into electrical and heat energy through the chemical reaction of fuel (mainly hydrogen) with oxidant (oxygen in the air) under certain conditions. They have the advantages of high efficiency, little pollution, low noise, high power density, and modular structure. Because of these outstanding advantages, fuel cells generating technology is widely used in vehicles power generation systems, fixed power generating systems and portable power, other systems, etc.

Fuel cells have been in demonstrated applications for years in many countries, which provide adequate evidence for the commercialization of fuel cells. But at present the fuel cells have not been fully industrialized partly because the fuel cells are very expensive, and partly because the technology of fuel cells is immature at the moment.

In today's world, when the energy crisis and environmental pollution are getting increasing attention, fuel cell technology is considered as the preferred clean and efficient power generation technology in the twenty-first century. Governments of different countries and large companies have invested heavily on the fuel cell technology research and development. The U.S. government takes the fuel cell technology as one of the key technologies to maintain the U.S. economic prosperity and national security. The Japanese Government gave strong support to the fuel cell technology research in the "Project Moonlight" and "Project Sunshine." The Canadian government determines the fuel cell the first choice in terms of the

Fig. 2.52 The household fuel cell displayed in Japan Pavilion

clean energy system in the twenty-first century. In Germany and Korea, fuel cell technology has been considered as one of the key power generation technologies in the future. China government has also paid much attention to research of fuel cells. The study of hydrogen energy and fuel cell technology has been listed as one of the major cutting-edge technologies in "Outline on National Planning for Medium and Long-term Scientific and Technological Development (2006–2010)." As the problems in fuel cell technology are gradually solved and the cost reduced, fuel cell will go to commercialization soon. In short, fuel cell is going from the stage of "research–development—demonstration" to the stage of commercialization.

2.1.4.2 Review of the Fuel Cell Applications in the Expo

Fuel cells are widely used at present, mainly reflected in the application of vehicles. In Expo 2010, there were 196 vehicles which used the fuel cell to provide energy, including 6 fuel cell buses, 90 fuel cell cars, 100 fuel cell sightseeing buses. To support the running of new energy vehicles, one hydrogen refueling station was built inside the Expo site and one recharging line was added on the Expo avenue to charge the buses of electric capacity; one hydrogen replenishing station and two moving hydrogen refueling vehicles were built around the Expo site near Jiyang Road, refueling the 50 fuel cell cars, 6 fuel cell buses, and 100 fuel cell tour buses.

At the Expo, a green household fuel cell named ENE-FARM from Panasonic (shown as Fig. 2.52) was displayed in Japan Pavilion. Launched in 2009, the product has the output power of 1 kW, providing 60 % household electricity for the average family, reducing 37 % of the carbon dioxide emission (compared with thermal power generation), ensuring a minimum 40,000-h running, and durable enough to last about 10 years. Home fuel cells have a promising market, it is

Fig. 2.53 Fuel cell model in Airbus A320 aircraft

estimated that in 2010, about 60,000 sets were sold, with the value of 650 million US dollars. On January 28, 2009, six Japanese companies namely, Tokyo Gas, Osaka Gas, Toho Gas, Western Gas, Nippon Oil, and Astomos Energy jointly established a union called ENE-FARM (energy farms) with the common purpose as "to build a nation on environment," proposed a slogan of "working towards the popularization of home fuel cell system for carbon dioxide emission reduction," announced that in 2009 the home fuel cell cogeneration system production program would be launched, with the first batch as 5,000 units, and expected to achieve a goal of selling 2.5 million units by 2030, equivalent to reducing 30 million tons of carbon dioxide emissions, about 1 year's carbon dioxide absorption by 5,600 km^2 forests. The ENE-FARM program received a strong support from the Japanese government—each ENE-FARM home fuel cell system enjoyed 45 % of the government subsidy (1.4 million Japanese Yen).

In the Hamburg Home Pavilion, people can see a fuel cell model used in Airbus A320 aircraft in the proportion of 1:20 (shown as Fig. 2.53). This is an important part of the general plan that Airbus is committed to the environmental aviation industry construction.

2.1.4.3 Proposal for Fuel Cell Promotion

Being an efficient, clean power generation technology, fuel cell technology is the hot and frontier technology in international energy research and development, which can solve the contradiction between the power generation efficiency of fossil fuel and environmental pollution, thus it has great potential of development and wild prospects of application. The promotion and application of fuel cells will provide huge economic and social benefits.

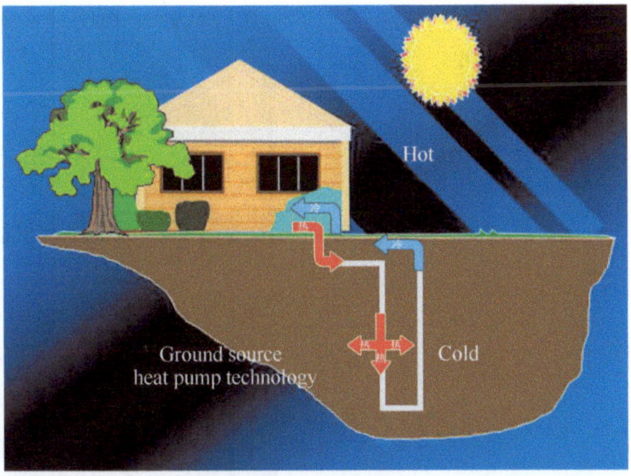

Fig. 2.54 Working principle diagram of ground source heat pump

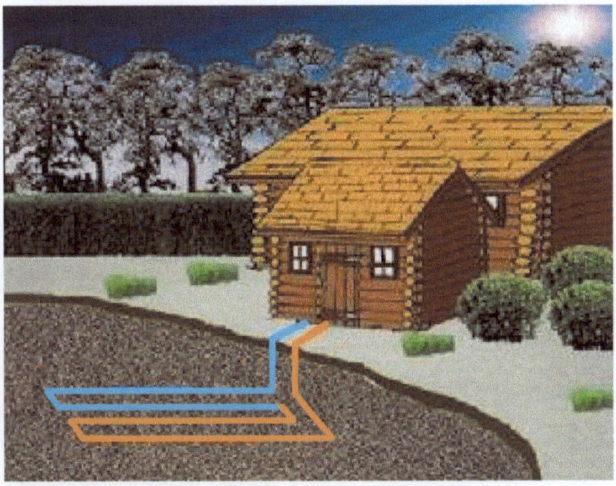

Fig. 2.55 Horizontal ground source heat pump

2.1.5 Geothermal Energy

2.1.5.1 Ground Source Heat Pump

Ground source heat pump (GSHP) is a central heating and/or cooling system that uses the water source on the earth's surface (groundwater, river, and lakes) and the solar energy and geothermal energy absorbed in the soil. It is an efficient and energy-saving air-conditioning system.

GSHP conveys a small amount of high quality energy (electrical energy) and uses a heat pump to force the transfer of thermal energy from low temperature to high temperature. Geothermal energy can be used as both the heat source in winter and cooling source in summer. That is, in winter, the stored energy in the ground is "extracted" to heat the room; in summer, the indoor heat is transferred back into the

Fig. 2.56 Vertical ground source heat pump

Fig. 2.57 Pond/lake ground source heat pump

ground. Normally, when geothermal pump consumes 1 kW energy, over 4 kW energy heat or cool energy will be obtained. Figure 2.54 shows the working principle diagram of the GSHP.

There are four basic types of GSHPs: horizontal, vertical, pond/lake (closed-loop systems) and underground water type (open-loop option) (See Figs. 2.55, 2.56, 2.57, and 2.58).

(1) **Horizontal ground source heat pump**: A horizontal closed loop system composed of pipes is buried horizontally 2 m underground (Figure 2.55), making heat exchange with soil. This type of system is the most common layout used in the buildings with small heating and cooling areas, such as villas and compact monomer buildings. It requires less initial investment and easy construction.

Fig. 2.58 Underground water GSHP

(2) **Vertical ground source heat pump:** A vertical closed loop field is composed of pipes that run vertically in the ground, typically 50–400 m deep in the rock-soil to have heat transfer with the soil (Fig. 2.56). This type of system is commonly used in the buildings with large heating or cooling areas and with certain space around, such as villas and office buildings. This system requires great initial investment and complicate techniques in construction.
(3) **Pond/lake ground source heat pump:** the system achieves heat exchange with the river, lake, and seawater through closed system lying at the bottom of the water (Fig. 2.57). This system is suitable for the building with mid- and small cooling and heating area and located close to the water.
(4) **Underground water source heat pump:** with the closed system of the unit, heat exchange is made between a heat exchanger and deep groundwater drawn by pumps (Fig. 2.58).

GSHP can save more than two-thirds of the electricity than electric boiler and more than half of the energy than fuel boilers. The heating temperature of GSHP is stable throughout the year, which is usually 10–25 °C. The coefficient of cooling and heat is up to 3.5–4.4 which saves plenty of energy and power. One set of the equipment only can meet the requirement of heating, cooling, domestic water supply, which reduces the initial investment of equipment. When GSHP is running, it does not consume water nor pollute water. Nor does it need any boilers, cooling towers or site for the waste fuel, which has significant environmental benefits. The power consumption of GSHP can be over 40 % less than air source heat pump and 70 % less than electric heating. Its efficiency of heating system is nearly 50 % higher on average than that of gas fired-boiler.

GSHP can be used in hotels, residential quarters, apartments, factories shopping malls, office buildings, schools, and other buildings. The small GSHP is more suitable for heating and air conditioning of villatic residences. The running costs are

Fig. 2.59 The water source heat pump unit

50–60 % higher than that of central air-conditioning. Over the past decades, especially in recent years GSHP system has made rapid development in North America such as USA, Canada, and some countries in northern Europe such as Switzerland, Sweden. China's GSHP market is gradually flourishing. It is estimated that this technology will become the most effective air conditioning technology for heating and cooling in the twenty-first century.

2.1.5.2 Water Source Heat Pump

Water source heat pump (WSHP), which is one or a group of systems, extracts the energy through water and provides heating and cooling. The WSHP unit is a device that extracts the low-grade heat from the surface water that cannot be used directly and then converts it into high-grade energy which can be used directly by consuming a small amount of high-grade energy. WSHP use solar energy and geothermal energy for cooling and heating, which is also regarded as one of the "GSHPs." After rigorous testing and calculations of the application examples of the heat pump in different areas, the heating COP of WSHP is estimated between 3.3 and 4.4 and the refrigerating COP is between 4.1 and 5.8. Figure 2.59 shows the WSHP unit.

The water source on the shallow surface of the earth (generally within 1,000 m depth), such as groundwater, surface rivers, lakes, and oceans absorbs considerable energy from the solar radiation and usually has a stable temperature. The working principle of the WSHP technology is that it achieves a transfer of low level heat energy to high level energy by imputing a small amount of high-grade energy (e.g. electric energy). It uses the water as a heat source in the winter or a heat sink heat in the summer. That is to say, heat is "taken out" from the building in summer and then released into the water. Since the water temperature is low, the heat can be removed efficiently to achieve cooling inside the building in summer. In winter, it is the WSHP unit that "extracts" heat from the water to the building. The working principle of WSHP is shown in Fig. 2.60.

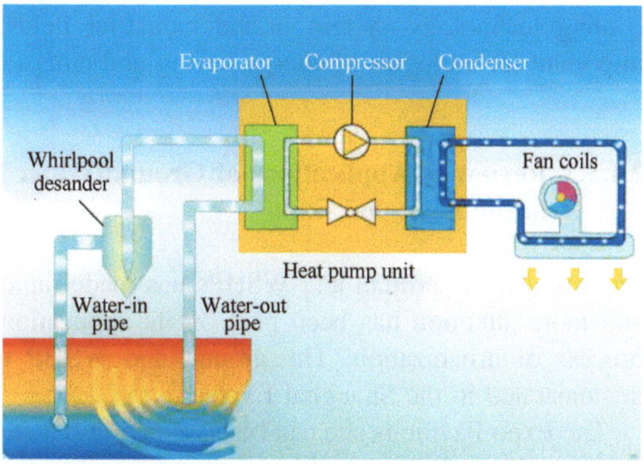

Fig. 2.60 The working principle diagram of water source heat pump

WSHP has significant economic and environmental benefits, as in the following aspects:

(1) **Environmental friendly**: WSHP is an energy conversion air conditioning system by using surface water as the heat source (in the winter) and heat sink (in the summer). When heating, no need for installing the boiler room system, including coal, gas, and oil, with no combustion process, which avoids smoke, sewage, and other pollution. When cooling, the pump does not need cooling tower so there is no risk of noise, mold pollution or water loss caused by the cooling tower. Therefore, WSHP is a technology using clean renewable energy.
(2) **High Efficient**: In winter, water temperature used by WSHP unit is 12–21 °C, higher than the ambient air temperature, so the evaporation temperature of heat pump cycle will rise and energy efficiency is also improved. In summer, The temperature of water is 18–35 °C, lower than the ambient air temperature, so the refrigeration condensing temperature decreases and the cooling effect is better than air-cooled and cooling tower type and the efficiency is improved as well. According to U.S. Environmental Protection Agency (EPA), WSHP with proper design and installation can save an average of 30–40 % of the heating and cooling air conditioning running costs for users.
(3) **Energy Saving**: The energy itself used by WSHPs is a clean energy, but in power generation, it consumes primary energy and leads to emissions of pollutants and greenhouse gas. Well-designed WSHP unit can reduce over 30 % electricity consumption compared to air source heat pumps and over 70 % electricity consumption compared to electric heating. Therefore, the WSHP can save energy as well as reduce pollution emission and greenhouse effect in the process of primary energy consumption.

WSHP technology is not only being widely used in all the central air-conditioning system of civil buildings, public buildings, military construction that need heating, cooling, and hot water, but also applied in the frozen, refrigeration, and cooling technology system in the industrial fields, which is one of the most important technologies of energy saving and emission reduction.

2.1.5.3 Review of Applications of Ground/Water Source Heat Pump in the Shanghai Expo

The research on ground and WSHPs has made rapid development in recent years and more attention has been paid on its applications in China, especially in the process of urbanization. The ground and WSHP technologies have been fully demonstrated in the Shanghai Expo.

The Expo Pavilions site can be classified into three kinds based on the supply of heating and cooling, namely, permanent pavilions with both heating and cooling sources, temporary pavilions with only cooling sources and commercial service facilities that distribute heating and cooling.

(1) **Applications of ground source/ water source heat pump in the Expo**

River water and GSHP cooling/heating system are mainly used in the 3.22 km^2 covering five functional areas within the Expo fence, to ease the "heat island" effect in parts of the city. The landmark buildings in Shanghai Expo—the World Expo Axis, the Expo Center, Performing Arts Center and the Pavilion of City Future—have adopted such technology.

Taking geographic conditions and surrounding environment into account, closed-loop GSHP, the river WSHP (from Huangpu River) and ice storage system are the main cooling and heating methods used in the permanent pavilions. The main applied Pavilions are as follows:

(1) Expo Axis: Base area of 130,000 m^2, total construction area of 248,000 m^2, 7 closed-loop GSHPs + 8 river WSHP units.
(2) Expo Center: Total construction area of 141990 m^2, 3 river water source heat pumps + 2 duplex status screw ice storage units + 2 gas hot water boilers.
(3) Expo Cultural Center: Total construction area of 140,277 m^2, 2 river water source heat pumps + 3 duplex state centrifugal ice storage units + 2 gas-fired boiler units.
(4) China Pavilion: Total construction area of 153,000 m^2, 3 three-grade centrifugal cooling water units + 3 duplex state centrifugal ice storage units + 3 gas-fired boiler units.
(5) Theme Pavilion (Pavilion of Future, Pavilion of Urban Planet and Pavilion of Footprint): Total construction area of 152,316.5 m^2, 3 sets of water cooled centrifugal chiller variable frequency + 8 sets of air source Screw heat pumps.
(6) Other temporary pavilions applying ground/water source heat pump are: Hamburg House, London ZedPavilion, Shanghai Home and Iceland Pavilion.

Fig. 2.61 Hamburg house in the Expo

Fig. 2.62 Schematic diagram of the GSHP along the Expo Axis

The Hamburg House in Shanghai Expo (Fig. 2.61) uses under GSHP air conditioning system through the buried pipes, making the U-tube in close contact with the pile and the pile with the ground to reduce thermal contact resistance and strengthen heat transfer between the cycle fluid and the soil of the earth. This project uses a special type of 3U pipe laying, of which 2U for inlet water and 1U for backwater to ensure balance and efficiency of .heat transfer. This method is also known as "energy piles." Due to the special requirements of its construction process, it is rarely seen at home. The application of this technology in Hamburg House indicates a new direction for the domestic GSHP development. Capillary radiant cooling/heating is used indoors, which is the most comfortable way, with multi-control combination modes available: refrigeration + fresh air, cooling + heating, heating, cooling, and dehumidifying mode (heat recovery at the same time), ventilation mode, among which the natural cooling mode is a new energy-saving way derived from the conventional GSHP system technology. Meanwhile, the host machine of the GSHP has complete control functions: loading and unloading (host with high efficiency at full load, low load adaptability), waste heat recovery control,

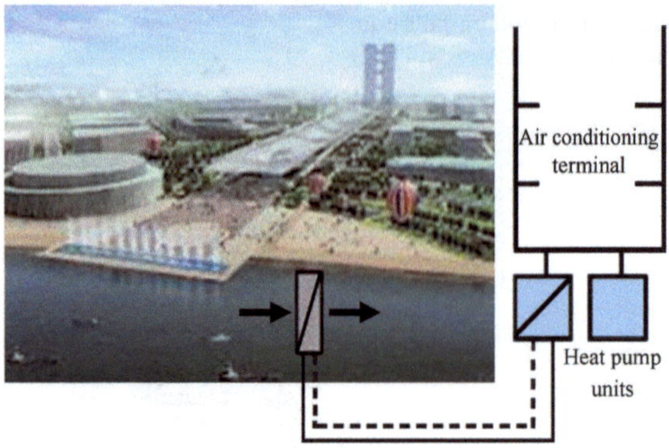

Fig. 2.63 Schematic diagram of the river WSHP along the Expo Axis

(the end, load, the source water) pump control, hot water pump, indoor temperature/humidity adjustment, water temperature control of the capillary side, anti-condensation, building automation and telecommunications connections, etc.

Take the Expo Axis as an example, near the end of the Huangpu River and the Dongjiadu Bend, a water intake was set up and then a heat pump was installed near the Expo Axis with about 1,200 tons of Huangpu River water running through the heat pump per hour which takes away the heat in the Expo Axis area and "exchange" the water of relatively lower temperature. The river water is discharged back to the river while the cool air is released through the different large and small channels. The river WSHP together with a small proportion of the GSHP will meet the cooling and heating demands of the Expo Axis. Figure 2.62 shows schematic diagram of the GSHP along the Expo Axis. Figure 2.63 shows schematic diagram of the river WSHP along the Expo Axis. Figure 2.64 shows joint design of Expo Axis River water and GSHPs.

In summer, river source heat pump system is used primarily, supplemented by the ground-source heat pump system, at the same time, the river water centrifugal chiller has the priority to be opened and used; in winter, it is the reverse, GSHP system is primarily used, supplemented by river WSHP system; when the river temperature is low, the suction flow of the river pump will be increased to ensure the thermal balance the of underground soil in winter and summer.

The engine room of the river WSHP system is located in the north of Expo Axis, concentrated in a room, which has five river WSHP units with cooling capacity of 1200 kW, heat capacity of 1100 kW, and three sets of river water centrifugal chillers with cooling capacity of 3800 kW. The utilization of the river water is of "direct style," which means that the river directly enters into the heat pump condenser or the evaporator and the condenser centrifugal chillers after being filtered. Its characteristics are: no heat exchangers, no heat change energy level loss, high operating efficiency; circulating water pumps with one level reduced leading to great increase of energy-saving efficiency. Before the river enters into the heat exchanger unit they have to go through two-stage filtration, first, two wide-meshed grid screens are used before water intake; then water goes through secondary filtration with automatic

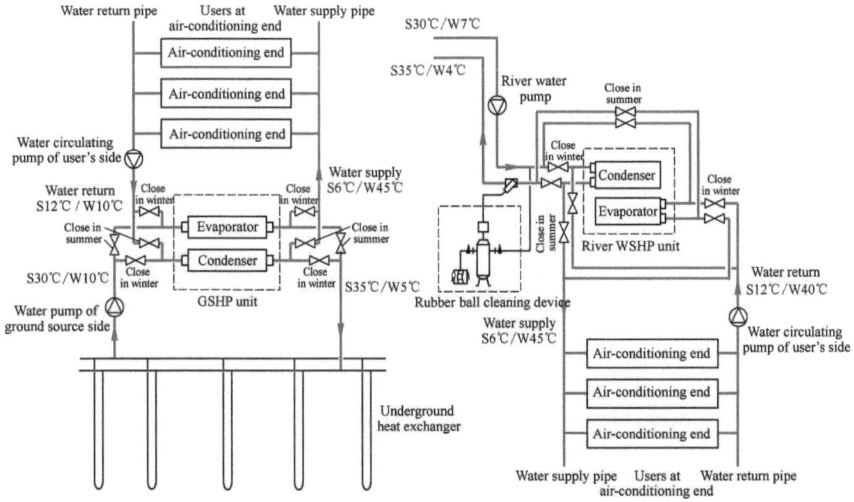

Fig. 2.64 Joint design of Expo Axis river water and ground source heat pump (*source* China Construction News)

backwash filter after it enters the room. The maximum designed amount of river is 6,000 m^3/h. In summer, the designed temperature of the intake water temperature is 30 °C and the drainage water temperature is 35 °C under the cooling condition; in winter, the designed temperature of the intake water temperature is 7 °C and drainage water temperature is 4 °C under the heating condition.

The transformation of the old Nanshi Power Plant has made full use of the technology of river WSHP system, which supplies cold and heat source for pavilions of about 15 million m^2 in Zone E of Puxi area during the Expo. and will continue exerting its role after the Expo, when the installed capacity will be further expanded to 41,379 kW, supplying cold and heat sources for the buildings of around 450,000 m^2 through the pipes ranging from 20 to 70 cm in diameter.

(2) **Recommendations on Promotion of the ground/water source heat pump**

The operating efficiency of GSHP is higher than the conventional air conditioning system by about 40–60 %. In addition, the characteristic of constant ground (water) temperature makes the heat pump more reliable, stable and the overall system maintenance costs is much lower than that of the boiler-refrigerating machine system, ensuring the system's efficiency and economy. Concerning its economy, it depends on many factors, such as different regions, geological conditions, energy structures and prices of energy etc., which will directly affect its economic benefit (Fig. 2.65). According to the experiences of foreign countries, due to the low operating cost of ground-source heat pump, the increased initial investment can be recovered in 3–7 years; the average cost of the ground-source heat pump system in the entire service period will be lower than that of the traditional air conditioning system. Among the Western countries, the United States makes the most intensive efforts in

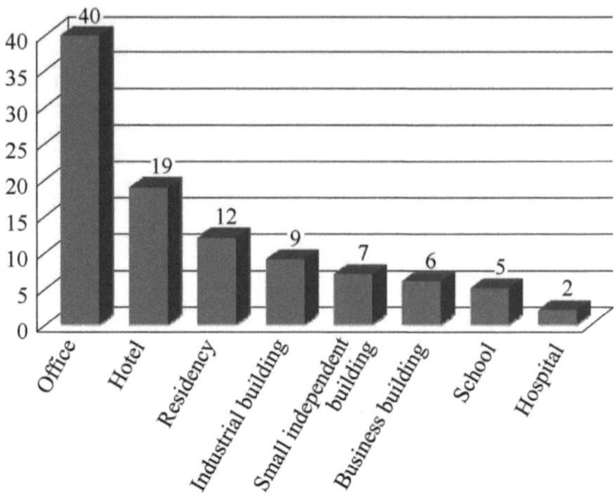

Fig. 2.65 Application proportion of GSHP technology in different type of buildings (%, survey of 160 projects in 2006)

Table 2.3 Installation of ground source heat pump in the world's major countries (2005)

Country	Total installed capacity (MW)	Annual output of heat (GWh/y)	Total installed number (units)
Austria	275	370	23,000
Germany	560	840	40,000
Canada	435	300	36,000
United States	3,730	3,720	500,000
Sweden	2,000	8,000	200,000
Switzerland	440	660	25,000

promoting GSHP, with all the data higher than those of the rest of the world in terms of the utilization scope. In 1997, 40,000 GSHPs of 12 kW were installed, while in 2000, the number reached 400,000. It is expected to reach 1.5 million in 2010. The main application areas are schools and office buildings. Table 2.3 shows the installation of GSHP in the world's major countries in 2005.

2.2 Energy Efficient Technologies

2.2.1 Building Energy Efficiency

2.2.1.1 Overview

Building Energy Efficiency (or Energy Saving in Buildings), was initially put forward in developed countries, meaning reducing energy loss in buildings, but the term now universally refers to "improving energy efficiency in buildings." That is, utilizing energy in a rational way and improve energy efficiency on conditions of ensuring the basic building functions and maintaining the comfort in buildings.

Specifically, Building Energy Efficiency means that during the process of the planning, design, construction (reconstruction and expansion), transformation and use of the building, energy saving standard should be carried out, energy-saving technologies, process, equipment, materials and products should be used. The purpose is to improve the insulation performance of the exterior protected construction (or building envelope) and the efficiency of the heating, cooling and HAVC system to enhance the operations management of the energy using system of the building, by using the renewable energy sources, reduce the energy consumption in heating consumed, air conditioning, lighting and hot water supplying under the premise of ensuring the quality of the indoor environment.

(1) **Significance of Building Energy Efficiency**

China is a developing country with a large population and a tremendous numbers of buildings as well. Every year the newly built housing area reaches up to 100 million m^2, surpassing the total annual construction area of all the developed countries. Building energy consumption has grown rapidly. Energy consumption of buildings refers to the use of energy in buildings, including heating, air conditioning, hot water, lighting, cooking, household appliances, elevators and other energy consumption, among which energy consumption of heating and air conditioning accounts for about 60–70 %. Of 100 million m^2 of existing buildings, only 1 % is the energy-saving buildings, The rest are high energy consumers in terms of exterior protected construction, heating and air conditioning system. Heating energy consumption per unit area equals to twice of that in the developed countries of similar latitude. That is because the heat preservation performance in the exterior protected construction in our country is poor and 67 % of heating energy have been wasted. The new buildings which could really be recognized as "energy efficient" buildings are less than 1 million m^2. The total amount of energy consumption in buildings accounts for more than 27 % of the total energy consumption in China, which is gradually approaching 30 %.

In regard of the material consumption, China has a higher consumption compared to the developed countries: steel consumption is 10–25 % higher, 80 kg cement more are needed for 1 m^3 of concrete but only 25 % of the sewage is recycled. To achieve a sustainable development of the national economy, the implementation of building energy efficiency is imperative and urgent. At present, severe energy waste in building is founded in our country. Energy consumption in building is growing faster than the production of energy. If the situation of this high-energy consuming construction is allowed to continue, it will be difficult for the country's energy production to meet the demands of such a high energy consumption, which leads to the organization of a large-scale energy-saving renovation of the old houses which will cost more manpower and material resources. To improve energy using efficiency in buildings will greatly ease the situation of the national energy shortage and promote the development of national economy. Therefore, energy efficiency in buildings is an important measure for implementing sustainable development strategies, realize the objectives of national energy planning and reducing greenhouse gas emissions to be in line with the global trends. Reducing the total energy consumption in the building is one approach to improve the building energy efficiency.

(2) **New Energy Efficient Technologies in Buildings**

The ideal energy-efficient building should meet the following three requirements based on the minimum energy consumption: firstly, control the solar radiation receiving or preventing in different seasons and different regions; secondly, keep the room comfort in different seasons; thirdly, get the room ventilated whenever needed. Currently, the ways to save energy in construction include minimizing non-renewable energy consumption to improve energy efficiency, reducing energy loss in building envelope and cutting the energy consumption of building facilities. In the above three aspects, high and innovative technologies play a crucial role. Of course, we can also use some traditional building energy-saving technologies but they can be used in modern buildings only after enough demonstration of advanced scientific tests and theoretical analysis.

(i) *Reduce energy consumption and improve energy efficiency*

In order to maintain the environmental quality of the living space, heating is needed to raise the indoor temperature in cold season and cooling is needed to reduce the indoor temperature in hot season. Humidity is needed when the environment is too dry and damp dehumidifier is needed when necessary. Ultimately, energy consumption is needed to achieve all of these. From the perspective of energy efficiency, it is necessary to increase the efficiency of heating/cooling system, which includes the efficiency of the device itself, the network transmission, the measurement of client and indoor environmental controlling devices. All of these require relevant industries to adopt high and new technologies in design, installation, operation, regulation in energy-saving system, equipment, materials, and other aspects of business management. For example, there are three new technologies to save energy in heating systems. ① By using computers, balanced valves and dedicated intelligent instruments, we can allocate the pipe network flow rationally, which not only improves the heating quality, but also saves energy. ② By installing heat distribution meter and the temperature regulating valve on the user's radiator, the users can control the energy consumption to obtain comfort and save energy as well. ③ By using the new thermal insulation material wrapping heating pipes to reduce heat loss in the pipeline. In recent years, low-temperature floor radiant technology has been proved to have good energy efficiency. It adopts cross-linked polyethylene (PEX) pipe as water pipe, spirally twisted in the ground in a double-way. In winter, it supplies low-temperature hot water (from geotherm, solar energy or various waste heat of low-temperature) to the pipe; in summer, cold water to the pipe can reduce the floor temperature (currently only used for heating). Compared with the radiator of thermal convection, this technology has many advantages of even distribution of indoor temperature, comfortable environment, energy saving, easy measurement and easy maintenance.

(2) **Reduce energy loss in exterior protected construction of the building**

Energy loss in the exterior protected construction of the buildings is mainly caused by three factors: (a) external walls; (b) doors and windows and (c) roofs. The energy efficient technologies in the above three parts draw much attention from the construction industry all over the world. The main development trend is to develop efficient and economical heat insulation materials and practical construction techniques so as to improve the insulation, heat-shielding performance and enclosed-safety of the exterior protected construction.

① Exterior wall energy-saving technology

With regard to energy-saving wall, traditionally, people used to increase the thickness of the walls with single heavy materials to achieve insulation, which can no longer meet the requirement of energy saving and environment protection. The compound wall is becoming the mainstream. Compound wall usually uses block material reinforced concrete as the bear structure, combined with thermal insulation material, or integrating thin wall material with thermal insulation material as the wall in the frame structure. Currently building insulation materials include rock wool, slag wool, glass wool, polystyrene foam, expanded perlite, expanded vermiculite, aerated concrete and polystyrene slurry and other particles. The production of these materials requires special techniques and equipment, which is far beyond the reach of the traditional technology. It is worth mentioning one material called the polystyrene granule mortar, mixed by rubber powder, polystyrene granule lightweight aggregate material and water, plastering on the outer surface of the wall to form a non-cavity insulation layer. Polystyrene granule aggregate uses recycled waste polystyrene board after milled, and rubber powder itself is blended with a lot of fly ash, so this material is waste recycling and energy efficient. There are three kinds of composite wall insulation technologies: inner insulation layer, outer insulation layer and sandwich layer. The third way is widely used in China; In European countries, the common practice is installing the wall with foam polystyrene boards outside. In Germany, the external insulation buildings accounted for 80 % of the total, 70 % of which uses polystyrene foam board. Figure 2.66 shows the external wall insulation system.

② Energy-efficient windows and doors

Doors and windows not only have functions of lighting, ventilation and containment, but also play important roles in art of architectural processing. However, windows and doors are the parts where energy is most likely lost. To enlarge the lighting and ventilation area or highlight the characteristics of the modern architecture, the area of doors and windows of the buildings are designed larger and larger. Sometimes all-glass curtain wall is used, which raises a higher requirement for the structure of external maintenance in energy saving. Currently, the major ways of improving energy efficiency of windows and doors mainly focus on improving the insulation performance of the materials and enhancing the air-sealing property of doors and windows. As for the window materials, some high-tech

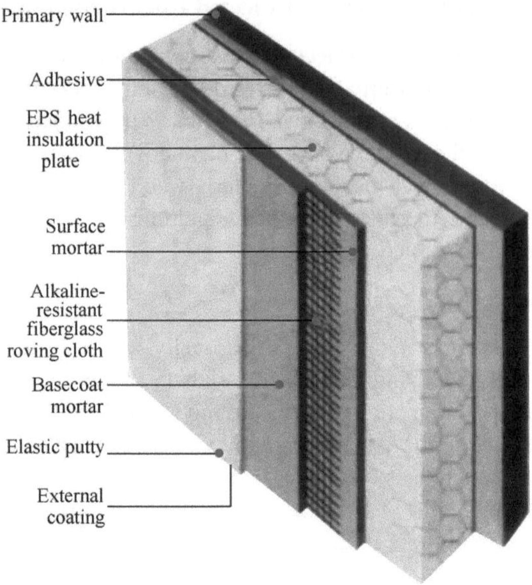

Fig. 2.66 External wall insulation system

energy-saving products have been on the market these years, such as insulating aluminum alloy profile, aluminum wood composite profiles, steel overall profile extrusion, plastic-wood composite profiles, and UPVC plastic profiles. Among them, UPVC plastic profiles is most widely used, which uses polymer materials—rigid PVC as its raw materials. This material has the advantages of not only less energy consumption, no pollution in the production process, but also small thermal conductivity and good sealing with multi-cavity structure and therefore is known to have good thermal insulation properties. UPVC plastic windows and doors have been used in European countries for many years, accounting for 50 % of the total plastic doors and windows installed in Germany. Since 1990s, the amount of plastic doors and windows used in our country has kept increasing, gradually replacing the steel, aluminum alloy and other materials which consumes much energy. To solve the problem of excessive energy loss caused by large-area glass, people use new and advanced technologies to process the ordinary glass into the hollow glass, coated glass (including reflective glass and heat-absorbing glass), and LOW-E high-strength fire-resistant glass (high strength low-E coating film fire-resistant glass), and metallic silver layer glass plated with vacuum magnetron sputtering and most extraordinary smart glass. Smart glass can sense the changes of light in the outside and respond to them with two outcomes: one is photochromic glass, which will turn dark when it senses the light radiation so the light can not penetrate it easily. Once the light irradiation stops the glass will become transparent again so light can penetrate it. Especially, when facing strong sunlight, the glass can block solar heat, while on overcast days, it brightens allowing the sunlight to enter the room.

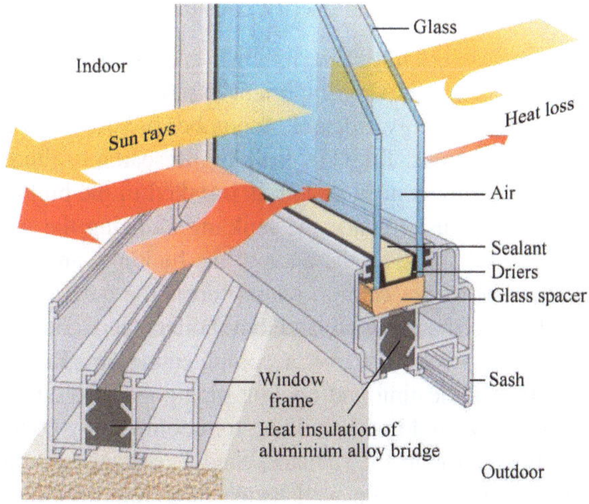

Fig. 2.67 Energy-efficient windows and doors

The other is electrochromic glass which contains two pieces of glass coated with conductive film and color material. By adjusting the voltage, change the color of the photochromic material to change the incoming sunlight (but due to its high production costs, it cannot go into practical use), all these glasses have good energy saving effects. Figure 2.67 shows the energy-efficient windows and doors.

③ Energy-saving roofs

Roof heat preservation and insulation is one of the key points of the energy-saving in the building envelope. In cold regions, roofs are set with insulation layer to prevent the loss of the indoor heat; in hot regions, cool layer of insulation is set on the roof to prevent the sun's radiation from going through the roof into the house; in regions where it is hot in summer and cold in winter (Yellow River to the Yangtze River region), it should give consideration to both winter and summer in regard of building energy efficiency. The technical measure commonly used in insulation is to install lightweight materials with smaller thermal conductivity under the roof waterproofing layer for insulation, such as perlite, glass wool, etc., or to install polystyrene foam above the waterproof layer of the roof. In the UK there is another approach to make insulation layer, that is, making use of recycled waste paper fiber. This paper fiber consumes little energy and has good insulation performance. After borax retardant treatment, this paper fiber can be fireproof. When in construction, people first make the sandwich layer on the roof with nails and then blow the paper fibers in between the sandwich to form insulating layer. Methods of roof insulation cooling include overhead ventilation, roof water storage or regular water spray, roof greening and so on. The above approaches can meet the requirements of energy-saving of the roof to varying degrees, but the most popular and recommended one is the use of smart technology and eco-technology to achieve the building energy efficiency, such as solar collector roof and controllable ventilation roof etc.

④ Reduction of the energy consumption of building facilities

Heating, cooling and lighting are the major energy consuming elements in buildings. So it is very important to reduce the energy consumption of these parts to achieve the energy efficient. A number of successful techniques in this regard are of great value to us. Take UK Building Research Institute (BRE) for example: advanced energy saving control system is installed in the building envelope, while transparent sandwich is used inside the building allowing the natural ventilation. Air comes in through the grille on the back of the building, and exhausts through the grille in the wall on the front ceiling, which forms the natural ventilation through the building. High-performance cold and hot boilers and conventional boilers are used interchangeably controlled by a computer system. Room temperature can be adjusted through the heating and cooling piping systems embedded under the floor. Another technology used in the building for cooling is by pouring cold water into pipes under the floor through the radiator: Step 1, extract the cold water from the ground with pumps from the deep well under the garage to the radiator; Step 2, pour the water back through another recharging well near the building. In order to reduce artificial lighting of the building, a comprehensive daylighting and lighting combination system is used, which is controlled by the building management system; Each unit of the building can receive sunlight where the system can be controlled by users and administrator through monitor. In the 100-seat lecture hall, two forms of lighting systems are installed which guarantees the brightness by using energy-saving fluorescent lamps and incandescent lamps to allow every audience to enjoy the same good visual effects and agreeable temperature.

(3) **Development of new energy-efficient building materials**

① Exterior insulation and finishing system (EIFS)

EIFS is a type of building exterior wall cladding system that provides exterior walls with an insulated finished surface and waterproofing in an integrated composite material system. This system came into being in the late 1970s when the last energy crisis occured. It was initially used in commercial buildings and later applied in civil buildings. Today, EIFS system accounts for 17.0 % of use in the commercial building walls, 3.5 % of civil building walls which grows at the rate of 17.0–18.0 % in civilian building walls per year. The system is a multi-layer external wall insulation system which contains the following parts. The main part is made of polystyrene foam insulation board, with the thickness of 30–120 mm. This part is fixed on the external wall by the means of synthetic adhesive or mechanical technology. The middle part is a long-lasting and waterproof primary polymer mortar. This basic layer is mainly used for insulation board to enhance and convey the external forces with glass fiber network. A final topcoat, or finish, is a beautiful and durable surface coating which is a colored, textured paint-like material. In order to prevent fading and cracking the covering material commonly adopts the technology of acrylic acid copolymer coating. A variety of colors and textures are available with strong durability and corrosion resistance.

② Structural Insulated Panels (SIPS)

Applied in both civil and commercial buildings, Structural Insulated Panels (SIPS) are a high performance building material as wall, floor and roof. The core of the board is a polystyrene foam sheet or Poly urethane foam sandwich layer, with a thickness of 120–240 mm. Different flat surfaces can be used on both sides as necessary. For example, the plywood type wood products can be used on both sides in construction projects. Buildings using SIPS will have higher strength, tighter building envelope, and better insulating properties, leading to shorter construction period and low construction cost. Moreover, SIPs-constructed building is energy saving and environmentally friendly. SIPS tend to come in sizes of 1.2 m in width with a maximum of 8 m. There are a series of dimensions for this material. Many factories can also customize the sizes of the product as required by the engineering. Contractors can just assembly this system at the construction site.

③ Insulated Concrete Foams (ICFS)

This product is an insulation template system made of recycled polystyrene foams and cement gelled materials, used for pouring in place the concrete walls or the foundation. In construction reinforcement bars are set within the template horizontally or vertically. When the wall is completed, the insulation template will serve as part of the permanent wall, formatting the insulation concrete walls both externally and internally. The template material applied externally on concrete walls meets the requirements of heat preservation, sound insulation and fire protection of the exterior wall.

2.2.1.2 Application of Building Energy Efficient Technologies in Expo 2010

Building energy efficiency refers to the use of new thermal insulation wall materials and heating air conditioning system of high energy efficient ratio in the construction under the premise of ensuring comfortable living, so as to save energy, reduce energy consumption and improve energy utilization efficiency. It involves three aspects such as energy efficiency in architectural materials, building structures and devices.

Building energy-efficient technologies have been widely used in Pavilions of the Expo site. For example, transparent concrete was applied on the external wall of the Italy Pavilion through the deployment of different proportions of the material allowing light to filter through, and thus reducing the use of lamps; German Pavilion used a meshy, breathable architectural fabric on its external wall which can reflect solar radiation to provide shading for the pavilion; In Japan Pavilion, environmental control technology was adopted in the pavilion design, making full use of the light, water, air and other natural resources. The double-layer membrane of high transmittance outside the pavilion light coupled with inner solar cells can take advantage of solar energy resources and realize a high efficient light guiding

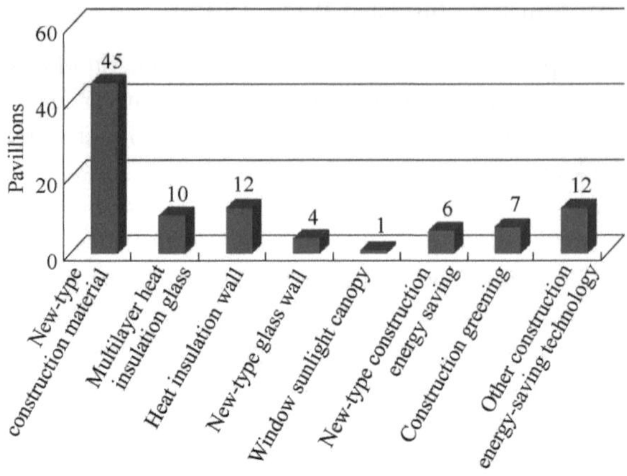

Fig. 2.68 Items of building energy-saving technologies applied in Expo site

and power generation. Inside the pavilion the latest technology of circular breathing pores was used. In terms of structure, Japan Pavilion uses an integrated semi-circular light-type structure combining roof with external wall, resulting in less impact on the surrounding environment in construction.

Figure 2.68 shows the items and numbers of energy-saving technologies applied in the Expo site, based on the comprehensive analysis.

(1) **Building Energy Efficient Materials**

The building energy efficient materials used in the Expo site include new thermal insulation materials, new plates, etc. The new thermal insulation materials include multilayer heat-insulating glass doors and windows, heat-insulating walls and heat-preserving paints, etc.

(i) *Energy-Saving Glass Doors and Windows*

According to the statistics, all doors and windows used in the World Expo pavilions are made of energy-saving glass. That is to say, the utilization rate of energy-saving glass has reached a full 100 %. Specifically, the types of energy-saving glass used are laminated heat-insulating glass, LOW-E hollow glass, coated glass, ultra white LOW-E glass, sodium metasilicate curtain wall and solar energy glass curtain wall, etc.

China Pavilion uses the LOW-E hollow laminated toughened glass supplied by Xinyi Glass Co. Ltd. As a new type of energy-saving safety glass, it can block most of the heat and ultraviolet in the sunshine, and has excellent performance in sound proofing and safety. The dimension of the LOW-E energy-saving glass in China Pavilion is larger than that of the common energy-saving building glass by 30 %, and its thickness increased by nearly 50 %, which almost reaches the limitation of the sophisticated equipment processing in the world.

The Low-E coated glass (Planibel-G®) offered by Asahi Special Colored Glaze Ltd is applied in Australia Pavilion, Brazil Pavilion, Spain Pavilion, Malaysia Pavilion, and Japan Pavilion, etc. Low-E coated glass is produced in the float

Fig. 2.69 Highly energy-saving doors and windows in Hamburg Pavilion

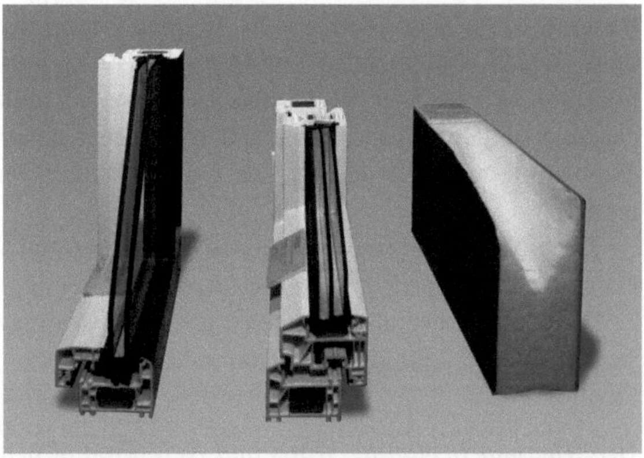

Fig. 2.70 The three-layered window structure exhibited in Broad Pavilion

Fig. 2.71 Heat-preserving walls exhibited in Broad Pavilion

processing line, When the glass is still under the high temperature, a layer of low-radioactive metal oxide film is formed on the surface of the glass by using the chemical vapor deposition technology. After annealing, this layer stick to the glass firmly with very good chemical stability, thermal stability and hardness. In freezing winter, the Low-E low-radioactive coat can reflect the heat absorbed by the glass back to the indoor to reduce the heat loss to the outdoor, and thus keeps the indoor room warm. While in hot summer, the Low-E low-radioactive coat can reflect the thermal radiation of the outdoor. In this way, the heat transmission from outdoor to indoor will be reduced so as to keep the indoor cool. Italy pavilion uses the new Low-E energy-saving glass, US pavilion chooses hollow glass made of Low-E coated glass from PPG.

For some other pavilions like the World Cultural Centre, Belgium Pavilion, UAE pavilion, Pavilion of Future, Power Plant in South City and Urban Best Practice Case Pavilion. They chose the low radioactive energy-saving hollow glass of Nanbo Group, which occupied over 70,000 m^2.

Besides, the Expo Axis, Theme Pavilion, and Performance Center have chosen the ultra-white glass produced by Jingjing Group. It is a high transparent low-iron glass, which is also called low-iron glass or high-transparent glass. With a luminousness rate of over 91 %, this new type of multifunctional high-end glass is characterized by its glittering and translucent design along with its elegant and exquisite taste, renowned as “the Crystal Prince” among the members of glasses family.

The ENERsign highly energy-saving doors and windows system displayed in Hamburg Pavilion (Fig. 2.69) has a window sealing property of 0.65 W/(m^2)K.

Hamburg House also displayed a kind of sealing material—Vac organ silicone sealant. Figure 2.70 shows the three-layered window structure exhibited in Broad Pavilion.

(ii) *Thermal Insulating Walls and New Type of insulating Paint*

Thermal insulating walls are adopted in most of the pavilions in the Expo Site to reduce the energy consumption of buildings. Figure 2.71 shows the thermal-insulating wall in Broad Pavilion. Figure 2.72 shows the insulation materials in Broad Pavilion.

Norway Pavilion has applied ElastosprayTM polyurethane spraying foam of BASF to ensure the heat preservation of the roof.

(iii) *New type of plates*

New type of plates have been used extensively in the Expo venues. The exterior wall of the Vanke Pavilion applies the wheat straw planking pressed from wheat straw. It is an environmentally friendly material and has almost zero formaldehyde emission. It features in outstanding nail-holding ability, both inward and outward fineness, and good fire-resistant property, as shown in Fig. 2.73.

Finland Pavilion (with the exterior wall covered with scale-shaped plaque) and China Pavilion (in the third exhibition area) both apply the UPM ProFi wood-plastic composite material. It is a special material, integrating the advantages of

Fig. 2.72 Insulation materials in Broad Pavilion

Fig. 2.73 Wheat straw plate on the exterior wall of Vanke Pavilion

both wood fiber and plastic, made from reclaimable material, solid and moisture-proof. The raw material of UPM ProFi comes from the scrap material in the production of sticky label, with the main components as paper and plastic. Therefore, this kind of composite material can either be incinerated or recycled in the process of production. See Fig. 2.74.

Hamburg Pavilion has used the BASF's Neopor in its hollow wall, a heat preserving plate which can increase the heat-preserving rate by 20 %.

The exterior wall of the Portugal Pavilion applies cork wood from Portugal, which mainly features light-weight, large buoyancy, strong elasticity and compression resistance. Besides, it is moisture proof and abrasion proof, and has good impermeability, low conductivity, heat insulating and sound-proofing, and high insulativity. As it is not inflammable, it can slow down the spread of fire and will not lead to allergic reaction, etc. See Fig. 2.75.

The rattan lath outside the exterior wall of Spain Pavilion is sturdy and durable, and at the same time allowing the light and air penetrating in. See Fig. 2.76

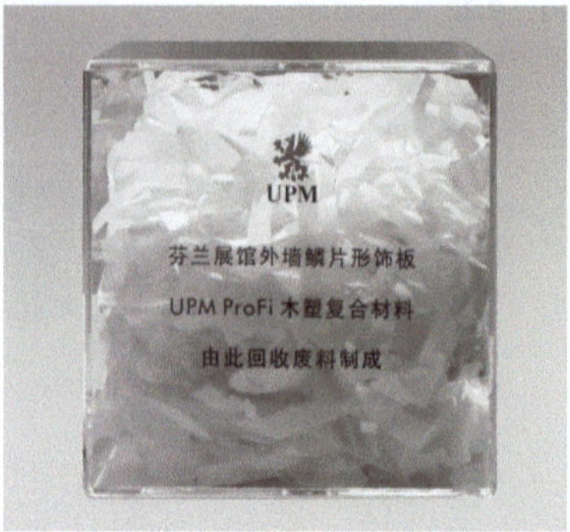

Fig. 2.74 Scale-shaped plaque on the exterior wall of Finland Pavilion

Fig. 2.75 Cork exterior wall of Portugal Pavilion

Fig. 2.76 Rattan lath exterior wall of Spain Pavilion

Fig. 2.77 Sun-shading system in Madrid Pavilion

Fig. 2.78 New energy sun-shading device in Shanghai Case Pavilion

(2) **Energy Conservation in Building Structure**

(i) *Sun-Shading System*

Sun-shading can effectively block the solar radiation and reduce the workload of air-conditioning in summer. The common sun-shading device includes cornice, visor and external venetian blinds, adjustable awning, external roller shutter, etc. Figure 2.77 shows the sun-shading system outside the Madrid Pavilion in Shanghai Expo, which is made of reeds through pressing and has a good performance in shading sunshine and preventing heat and economical and environmentally friendly as well.

The sun-shading system in Shanghai Case Pavilion can be classified into the following (Fig. 2.78): sun-shading through construction, structural sun-shading, greening sun-shading and sun-shading through new energy components.

Fig. 2.79 Southern afforested wall of Alsace Case Pavilion

(ii) *Construction greening*

The functions of construction greening are as follows:

a. Ease the heat-island effect: Planting trees or grass on the roof can effectively ease the heat island effect of the city, adjust ambient temperature in the neighboring cities and bring down the indoor temperature. In hot summer, the surface temperature under the vegetation layer on the roof usually ranges from 20 to 25 °C, which is superior to ordinary heat insulating method.
b. Save energy and protect environment: The green roof matrix has extremely strong water retention. When the roof is afforested, its draining strength can be reduced by 70 %, which not only saves water, but also alleviates the pressure of the city drainage system. Meanwhile, it can cut down the air-conditioning use frequency on the top floor, and thus save a lot more electricity.
c. Promote ecological balance: The roof greening campaign in urban buildings is a way to return the effective ecological area to the nature, perfect the ecological system, and increase the air humidity as well. It can also add to the city's oxygen content, alleviate air pollution and maintain the balance of carbon against oxygen in the urban air.
d. Beautify buildings and protect the roof: greening on the roof can beautify the city's environment and the living environment. Moreover, it can protect the building surface from aging or expanding and contracting caused by temperature difference, so as to prevent the roof from leakage caused by the diaclase.

The most typical building greening examples in the Expo site were displayed through London Case Pavilion, Alaska Pavilion (Fig. 2.79), India Pavilion (Fig. 2.80), China Pavilion, Theme Pavilion (Fig. 2.81), Canada Pavilion, Ireland Pavilion (Fig. 2.82), Singapore Pavilion, and Saudi Pavilion.

Fig. 2.80 Afforested roof of India Pavilion

Fig. 2.81 Vertical afforested wall of Theme Pavilion

Through a rainwater collecting system, India Pavilion uses the factory-processed reclaimed water for irrigation.

The ecological green wall of Theme Pavilion, as large as 5,000 m^2, can annually hold back 870 tons of dust, achieve 3,175 tons of carbon sequestration, reduce 96 tons of carbon dioxide emissions and saves electricity consumption of air-conditioning in the summer by 125,000 kWh, undoubtedly making itself the "green lung" of the Expo site. In the summer, through the greening exterior wall it blocks the thermal radiation and reduces heat conduction, when the temperature is above 34 °C, the air temperature around the green wall can be reduced to around 31–32 ° C, lower by 2–3 °C. Compared to the conventional curtain wall, the ecological green wall lowers its energy consumption by 40 %, resulting in the reduction in the air-conditioning workload by over 15 %. In the winter, the ecological green wall ensures the wall space receiving enough solar radiation heat, and meanwhile it forms an insulation layer on its surface, which reduces the energy consumption from heat supply by 10 %.

Fig. 2.82 Roof and surrounding afforestation of Ireland Pavilion

Fig. 2.83 Roof air caps in London Case Pavilion

(3) **Energy-saving Ventilation System**

In Shanghai World Expo, many different energy-saving ventilation systems have been applied in different pavilions, among which the air-refreshing equipment in Broad Pavilion and the roof air cap system in London Case Pavilion are typical examples.

A specially structured air cap has been used in London Case Pavilion (Fig. 2.83), which has the function of passive ventilation and heat reclamation.

An air-refreshing equipment with function of air purification has also been displayed in the Broad Pavilion. With 7 types, the equipment can process different levels of air quantities, ranging from 125 to 20,000 m^3/h; with the purification rate of fresh air reaching 99 % and heat recovery rate 90 %.The equipment works through either joinable ventilation pipe or straight blow, with the power consumption of 0.7 W/m^3. See Fig. 2.84.

Fig. 2.84 Heat-recovery air-refreshing equipment exhibited in Broad Pavilion

2.2.1.3 The Reality of Building Energy Efficiency in China

(1) **The current building energy efficiency level in China is not high, mainly for three reasons:**

① limited only to energy saving in the building envelope and equipment;
② Imperfection of energy efficiency standards system;
③ Building energy audit and energy consumption monitoring have to be improved.

(2) **Suggestions on the promotion of building energy efficient**

① Establish energy efficient management system based on energy consumption data of large public buildings;
② Conduct the existing energy efficient buildings innovation with the guide of Scientific Outlook on Development;
③ Develop green buildings;
④ Establish low-carbon demonstration cities.

2.2.2 Ice Storage Technology

2.2.2.1 Introduction to Ice Storage Technology

Ice storage technology is a series of technologies of storing cheap electricity during off peak hours at night and release the coolness by melting for use during subsequent daytime peak hours. The system works with refrigerator. Figure 2.85 shows the working principle of ice storage air conditioning system.

Ice storage technology can save the cost of air-conditioning equipment and reduce 30–50 % of the installed capacity of the chiller. For users, the operation costs will be reduced as great as 30–50 %. Besides, it can reduce air pollutants

Fig. 2.85 Working principle for ice storage air-conditioning system

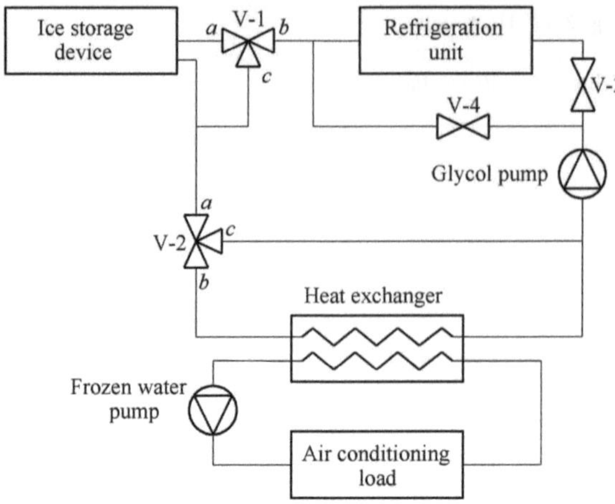

emission around the power plant as well as the CFC and combustion emissions around the buildings. Meanwhile, it can also enhance the power generation efficiency and thus improve energy utilization efficiency, which can achieve "load shifting" of the power, transferring peak power load and balance power supply.

As an efficient means of energy conservation in the twenty-first century, ice storage technology has a great potential for development and benefits mankind, resulting in a positive social effects and economic benefits. It is an important method to increase the electricity load of the grid, improve the overall benefit from investment in power utilities and reduce the emissions of carbon dioxide and sulfides, which is of great significance in environment protection.

2.2.2.2 Ice Storage Technology Utilization in the Shanghai Expo

Ice storage technology has been used in the China Pavilion, also known as "Oriental Crown." The working mechanism is very simple. That is, place a huge ice storage pool at the bottom of the pavilion, the refrigeration system could make ice during the off peak night time when the power rate is lower, while the ice will melt into cooling water to cool down the pavilion in the daytime. As Shanghai Expo was opening from May to October, covering summer and autumn, which required a large amount of cooling air. This technology can effectively help Shanghai to achieve the electricity "load shifting" during this period. According to a survey, by using the "Ice Storage" and "River Water Source Heat Pump" Technologies, the World Expo pavilions could reduce at least 15–18 % of the energy consumption compared with the conventional air conditioning system.

Apart from the utilization in China Pavilion, Ice storage Technology has also been applied in the permanent building "Four Pavilions along the Central Axis" in the Expo site. In the China Pavilion, 6 sets of Trane host were used, including 3 sets of 830 tons of cold duplex units, and 3 units of 800 tons of base load cooling unit.

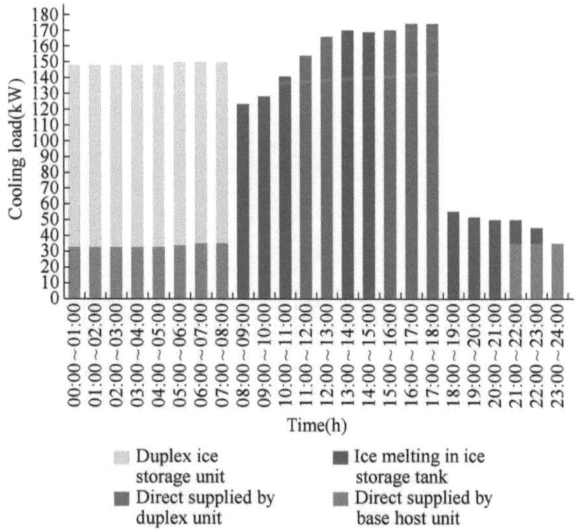

Fig. 2.86 Designed daily operation strategy during the Expo

The Art Performing Center have installed 3 sets of 650 tons of duplex status of the cold unit, two 650 tons of cold river WSHP unit and the end device. I

(1) The designed daily operation strategy during the Expo

For the host, maximum cooling storage is 928 kW, the daily ice melting amount is 923 kW. The airborne host, duplex status host and ice storage tank operated in the following five models, shown as Fig. 2.86.

(i) Duplex status host refrigeration mode (from 24:00 to 8:00 the next day), base host refrigeration;
(ii) Duplex status single mode of host refrigeration (11:00–13:00 and 15:00–18:00);
(iii) Ice-storage tank single mode of refrigeration (8:00–13:00 and 15:00–18:00);
(iv) Base host single mode of refrigeration (23:00–24:00);
(v) Base host + Ice-storage tank united mode of refrigeration (21:00–23:00).

(2) The daily operation strategy designed when no visits

Maximum cooling storage of the host is 928 kW, the daily ice melting amount is 910 kW. The airborne host, duplex status host and ice storage tank operated in the following three models. See Fig. 2.87.

(i) Duplex status host refrigeration mode (24:00–8:00 the next day), base host refrigeration;
(ii) Ice storage single refrigeration mode (8:00–23:00);
(iii) Base host + ice storage tank united refrigeration mode (21:00–23:00).

It is seen from the above Figs. 2.86 and 2.87 that large-scale units operate efficiently in full load during most of the day time, and the insufficient load of the

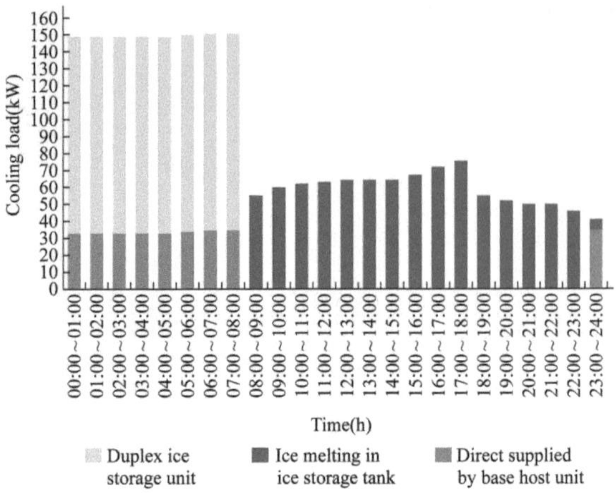

Fig. 2.87 The designed daily operation strategy with no visits

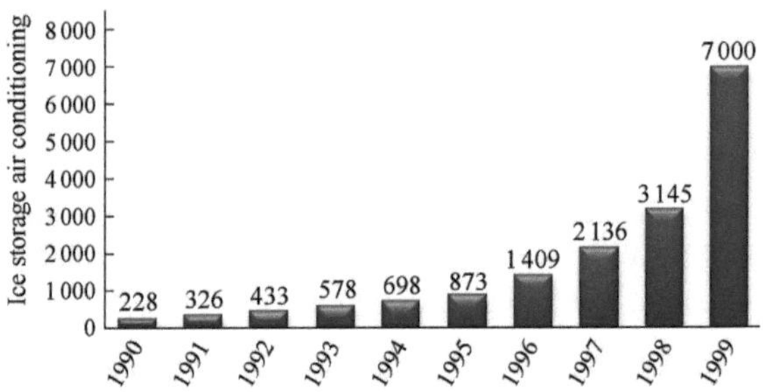

Fig. 2.88 Growth of ice storage air conditioning in Japan in 1990s

air-conditioner is adjusted by the ice storage system and thus increase the operation efficiency of the host while taking advantage of the off peak power rate.

2.2.2.3 Suggestions on the Ice Storage Technology

In the 1970s, due to the global energy crisis, plus the peak electric load growth in summer in the United States, Japan and some European industrial countries and the gap between peak and valley loads is widening at an alarming rate, resulting in a low-load operation of power plants at night. Thus ice storage technology was introduced to the centralized air conditioning system on an experiment basis. Since then the more and more ice storage air conditioning systems have been used in the engineering project. Take Japan for instance, which is among the fast growing countries, in 1990 it had only about 200 of ice storage air conditioning systems, in 1999 the number increased to 7,000 (see Fig. 2.88). In fact the ice storage air conditioning system development in the United States is also growing very fast (remarkably).

Currently the electricity departments in many cities in China have introduced the time-of-use (TOU) price structure and the relevant preferred policies, which will reduce the operating costs of users and thus encourage more people to use ice storage air-conditioners.

Ice storage air conditioning system features a higher technological content over conventional air conditioning system. We believe that it's an inevitable trend to develop ice-storage air conditioning system in China. So far there have been only 109 ice storage air conditioning systems in our country, of which more are in Zhejiang and Beijing, 32 and 27 respectively, accounting for 29.4 and 24.8 % of the total, as in other provinces and cities there are not more than 10 such systems. Table 2.4 shows cold storage air-conditioning applications in China.

2.2.3 Other Energy Saving Technologies

2.2.3.1 LED Technology

(1) White Light Emitting Diode (LED), also known as light-emitting diode, uses solid semiconductor chip as light-emitting material. When coupled with a forward voltage on both ends of the electrodes, the current carrier across the semiconductor will mix with each other and release excess energy in the form of photon and then emit light. Figure 2.89 is a diagram shows the **schematic diagram** of the light-emitting diode.

(i) *The Status of White LED development in China*

At present, many domestic enterprises have mastered the manufacturing technology of LED with small power chip. High power LED has high requirement of the epitaxial material. The leading Japanese companies in the industry strictly block the technology. Nichia, a Japanese company, who is known for its cutting-edge research in GaN, has worked with Toyota to develop a device of metal organic chemical vapor deposition (MOCVD). But they don't sell this equipment to foreign countries. Epitaxial wafer and chip manufacturing belong to the technology- and capital-intensive industries, so the entry barriers remain high. At present, the core patent technology is still in the firm grip of large companies, such as NICHIA of Japan, Cree of United States and Osram of German, etc. These companies rely on setting patents bulwark and filing patent litigation to keep other companies from entering the market to compete with them so as to obtain the high monopolistic market interests and form a strong patent alliance. Lack of core technology is a bottleneck in the development of LED in our country. According to statistics, by the end of 2009, China had obtained around 20,000 LED-related patents, but the vast majority confines to the downstream applications, packaging fields. Due to lack of core technology, the development of upstream and downstream development of LED is not balanced. The majority of enterprises are small with low-technology content with irrational industrial structure.

Table 2.4 Cold storage air-conditioning applications in China

Time	1993	1998	2001	2002	2003
Number of users	2	81	177	233	311

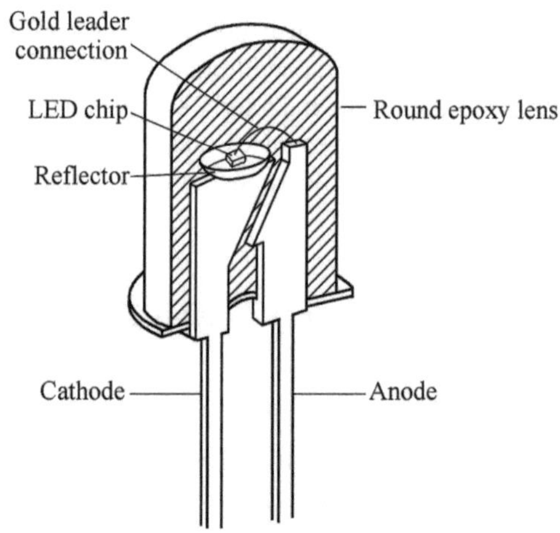

Fig. 2.89 Schematic diagram of light-emitting diode

The blue LED chip which can be put into volume production by major Chinese epitaxial wafer and chip manufacturing enterprises has the luminous efficiency of 110 lm/W (lumens/watt), while the luminous efficiency in the world has reached 140 lm/W.

(ii) *Advantages of LED in energy saving*

As a new solid-state light source, LED features small size, high light efficiency, long service time, quick response and high reliability. Now it has been recognized as a major lighting materials of energy saving in the world. Compared with traditional lighting technologies, the most important advantage of LED at present is that it is environmentally friendly and energy efficient.

Cao Dong, general manager of Shanghai Songer Lighting Engineering Co., Ltd, who is responsible for the LED Lighting project in the Expo Park and the theme Pavilion, told the reporters that according to accurate calculation, the 1,700 sets of LED lighting systems in the Expo Park could save 60 % energy compared with the traditional lighting, The white LED lighting system used in the arterial roads and stairs can save up to 75 % of the electricity.

(iii) *Application of monochromatic LED*

LED was first used as the light source for instrument indication. Later a variety of light-colored LED has been widely used in traffic lights and large display screens, producing great economic and social benefits. Take 12-inch red traffic

Fig. 2.90 Night scene of China Pavilion

lights for example, in the United States, people primarily used long life and low-efficiency 140-W incandescent lamp as a light source, which produced 2,000 lm of white light. After going through the red filter, 90 % of the light is lost and only 200 lm of red light is left. In the newly designed lights, however, Lumileds used 18 red LED light sources, including circuit losses, consuming a total power of 14 W but produced the same lighting effect.

Automotive signal lights are also important areas of LED light application. In 1987, China began to install brake light in cars because of the LED's quick response speed (nanosecond grade) which allows the drivers of trailing vehicles know the driving conditions earlier so that they can reduce rear-end accidents.

In addition, LED lights have been applied on red, green, blue full-color display in the outdoor and key chain-type mini flashlight and some other fields.

(2) **Applications of the new light source LED technology in the Expo**

According to the statistics, currently 1.3 billion pieces of LED chips have been used in the Expo Park, which appeared in indoor-and-outdoor lighting, landscape decoration, direction boards and information display screens. The whole Expo Park becomes the most concentrated area of semiconductor lighting technology demonstration and application. LED green light sources are applied in almost 80 % of the indoor lighting sources in the Expo pavilions, which can save almost 90 % of the electricity compared with incandescent lights.

LED has not only been applied to the whole landscape lighting of "Four Pavilions along One Axis" and the city's best practice area, but also used as the LED display screens to exhibit varied video contents, which is worth over 1 billion yuan. The LED lighting application on such a large scale in the Expo Park has a total energy-saving efficiency of over 70 %. LED creates an earthshaking scene with enormous visual impact and helps make a perfect Expo.

Over 2 million LED lights and 90,000 sets of full-colored lamps and light ribbons of different types were used along the 1 km long Expo Axis, which demonstrated the ideas of low-carbon Expo and green Expo.

Fig. 2.91 Night scene of Expo Axis

Fig. 2.92 Night scene of Shanghai Corporate Joint Pavilion

The LED lights also help creating a series of magnificent scenes, such as: the elegance of the Theme Pavilion, the dignity of the Expo Center, the future sense of the Cultural Center as well as the dynamic vigor of the Expo Axis and the shining of "the Chinese Red." Examples of LED applied in the Expo is seen in Figs. 2.90, 2.91, 2.92, and 2.93.

Take the Sun Valley of the Expo Axis for example, each intersection of the light-steel structure was installed with direct view LED luminous points, and there were some 80,000 such luminous points which can display images and characters with lower resolution through digital control system. Meanwhile, the large-sized tensile membrane on top of the Expo Boulevard also used LED for spot-lighting, which enabled the whole Expo Boulevard to make use of the dynamic changes of the LED full-spectrum to create a lively festive atmosphere.

More than 15,000 LED lights were arranged both inside and outside the "Magic Cube"—the theater. Inside the theater there were 100 spotlights, 100 ultraviolet lamp combinations, 75 digital mobile lights, together with 300 loudspeakers, Outside, 24 seven-kilowatt searchlights, 75 flashing lights, and 150 high-powered

Fig. 2.93 Night scene of Information and Communication Pavilion

loudspeakers were installed. These combinations together make the "Magic Cube" a breathtaking and dreamlike place.

In the Urban Best Practices Area located in Puxi Site, a section of road surface was paved with LED light tiles. When night fell, the LED system was showing different colors and patterns on the ground as time changes.

The "Ball of Energy" in the Energy Source Hall of Germany Pavilion was a great highlight, since the ball was installed with more than 400,000 light-emitting diodes. The ball would turn to one side where the cries are louder and emit dazzling light. In the huge sound wave on site, the Ball of Energy quickly rolled around, perfectly interacting with the audience.

The UK Pavilion at night was like a huge LED dandelion, as shown in Fig. 2.94, thousands of LED lights were "rooted" in some 60,000 Acrylic "tentacles," with each light guarding a seed. The image of the UK Pavilion thus became the favorite of many photographers.

LED, known as the "green light source of the twenty-first century," was used as the main lightening body in the Urban Best Practices Area, and it is the first time for LED to be applied in city blocks on a large scale. The creative designs such as "the singing floor," "ripples coming from standing on water," and "a plant that quickly flourishes with the increasing number of visitors," integrated highly effective, humanized and sustainable techniques, allowing the visitors to enjoy the beauty of the lights.

(3) **Recommendations on the new lighting source LED technology**

Nowadays, as "a Technology to Light the Future," LED has drawn the attention from more and more countries in the world. The development of the technology in the long-term will grow at a very high rate. It is estimated that the output value of LED all over the world has increased from 12.5 billion US dollars in 2004 to 50 billion US dollars by 2008. It has a huge market potential but faces growing competition as well.

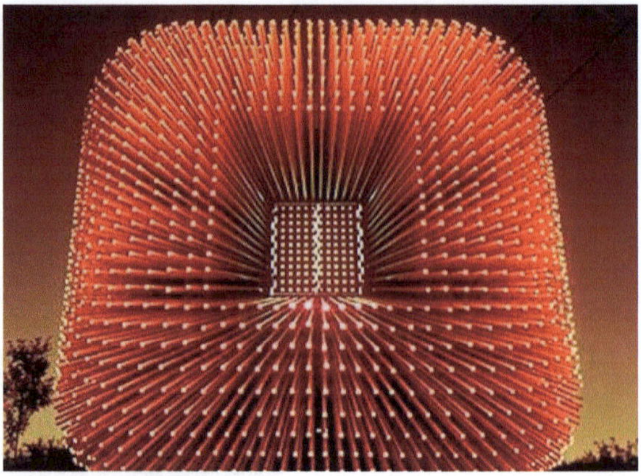

Fig. 2.94 UK Pavilion during the night

Table 2.5 Development target of the lighting industry in China

Technical index	LED (2010)	LED (2020)	Incandescent lighting	Fluorescent lighting
Luminous efficiency (lm/W)	100	200	20	75
Service life (kh)	50	100	4	10
Luminous flux (lm)	800	1,500	1,200	3,400
Input power (W)	8	7.5	75	40
Unit cost (Yuan)	33.2	24.9	2	14
Cost per mille lumens (Yuan)	41.5	16.6	1.7	4.1
Color Rendering Index (CRI) (Ra)	>80	>80	>95	75
Total cost per million lumens hourly (Yuan)	6.63	3.28	40	7.4

At present, China has become the most important production base of the middle- and down-stream LED products. It is estimated that the LED industry in China will reach 100 billion by the year 2010.

The technological innovation will maintain a sustainable development of LED. The impact of foreign technologies and the progress of domestic LED technologies will be the trend in the future 5–10 years. See Table 2.5, Figs. 2.95 and 2.96.

(i) *Actual status in the industry*

At present, the dominant technical route in producing white light LED both in domestic and overseas markets is still by using blue chip plus yellow phosphor. In the past decade, white LED manufacturing technology has achieved rapid development. Its main technical index, luminous efficiency, has increased by several times. For example, in the conditions of 20 mA, Nichia's laboratory products have reached 249 lm/W; at 350 mA, Cree's high-power laboratory products have reached

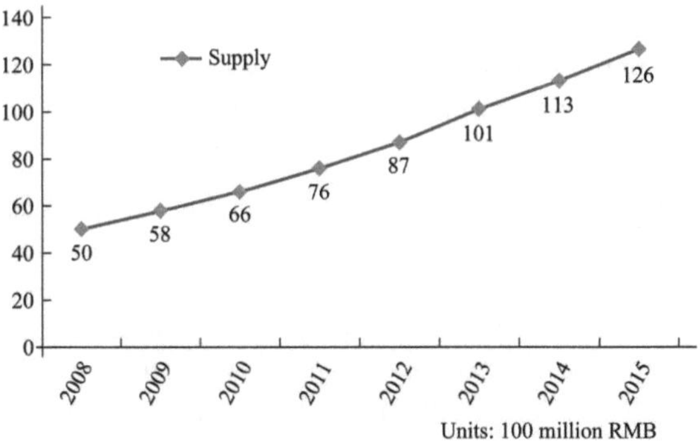

Fig. 2.95 High-power LED lighting products supply and forecast in China in 2008–2015

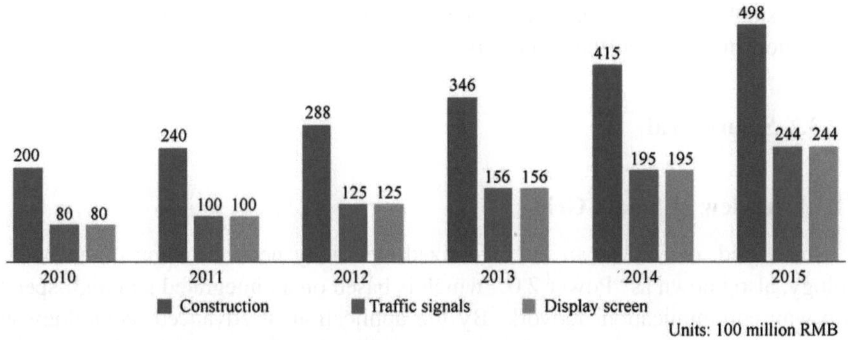

Fig. 2.96 High-power LED lighting market size forecasting in China in 2010–2015

208 lm/W. Cree's products in commercial production have reached 140 lm/W and are expected to reach 160 lm/W by the end of 2010. With the improvement of LED technology in overseas countries, the performance of domestic high-power white LED products has increased substantially. The luminous efficiency of the product in 2003 was only 30 lm/W, but reached 110 lm/W in 2010. In China, LED has been widely applied in many fields, with the luminous efficiency of white LED far surpassing that of incandescent and fluorescent lamps, close to that of high pressure sodium light and metal halide lights.

(ii) *Development trend*

White LED is an environmentally friendly, energy efficient and long life lighting products, with good prospects for development. The main restrictive factor in the white LED promotion is still t higher cost, e.g. a 7 W LED sphere bulb with the luminous flux same as that of a 40 W incandescent bulb has the manufacturing cost of about 50 yuan. Therefore, the main direction of white light LED development shall start with the basic aspects such as manufacturing equipment, dedicated materials, packaging technology, light fixture manufacturing and mass production in a bid to further improve the luminous efficiency and further reduce product costs, allowing users to save money while saving electricity.

(iii) *Suggestions for promotion*

a. The state shall formulate the guidance planning for industrial development and guide the LED lighting industry to develop in the right way.
b. Government of different levels shall stimulate the innovative enthusiasm of the enterprises with all kinds of projects of science and technology and encourage the joint development of production and research so as to make China's LED lighting technology keep up with the pace of world development.
c. The initial stage of any emerging strategic industries is inseparable from the government promotion, which requires governments of all levels to provide support through the demonstration project, product subsidies and such forms.
d. Establish relevant product standard to standardize and coordinate the relationship between the producers and the users.

2.2.3.2 Smart Grid

(1) **Overview of Smart Grid**

Smart grid, or SPG is an intelligentized electricity network using digital technology, also known as "Power 2.0," which is based on an integrated and high-speed two-way communication network. By the application of advanced technology in sensing and measurement, equipment, controlling methods and decision support system, it achieves the targets of being reliable, safe, economic, efficient, environmentally friendly and safe in use. Smart grid features self-healing, consumer motivation, attack resisting, higher quality power it provides to meet consumers' needs in the twenty-first century as well as power generation options which accommodate different forms of power generation connected to grids, besides, it enables electricity market to flourish and optimizes the assets while minimizing operations cost. The structural diagram of the smart grid is shown as Fig. 2.97.

Smart grid includes intelligent scheduling system that has priority to the access to clean energy, smart metering system with dynamic pricing and intelligent technology systems which optimizes load balancing by adjusting the power generation and electric equipment power. Power flows from the intensive generation stations to grid and from distribution network to the users. Meanwhile there are all sorts of forms of new and clean energies in the electric network, such as solar energy, wind energy, fuel cells and electric vehicles, etc. In addition, high-speed and two-way communication system achieve the information exchange between control center and network devices while advanced analytical tools and decision-making system ensure the security, stability and optimized operation of the smart grid. Figure 2.98 shows the overall planning of intelligent power distribution network.

Fig. 2.97 Smart grid structural diagram

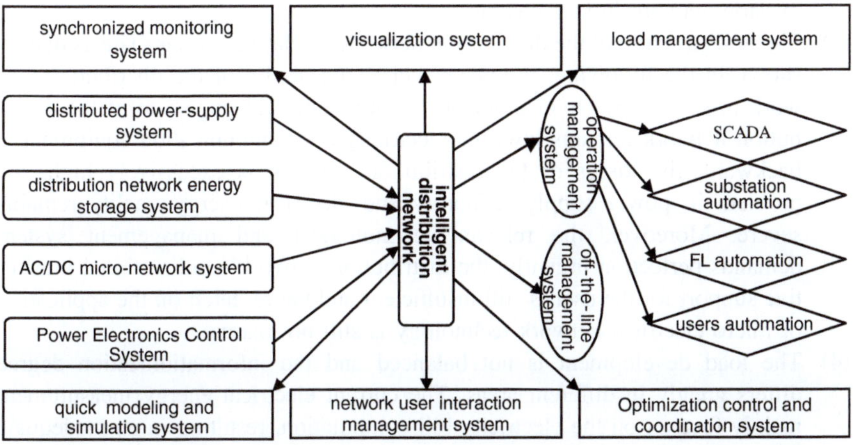

Fig. 2.98 Overall planning of intelligent power distribution network

Since the twenty-first century, the electric power construction in China has undergone rapid development: the installed capacity and electric energy production have been growing continuously and the electricity structure adjustment and technology upgrading have made certain achievements. With the power network scale expanding unceasingly, the power transmission capacity and voltage class being constantly enhanced and the network architecture strengthened, we have initially realized the aim of nationwide power networking. At the same time, we have also made great progress in the development of new energy generating technology. The project of upgrading urban and rural power grids has made remarkable achievement and the reform of the electricity market has achieved initial success. Nevertheless, there still exist the following problems:

(1) Although the power generating capacity has increased greatly, it still relies on the traditional thermal power as its energy source and the power supply

structure requires urgent adjustment and optimization, which is mainly manifested in the following factors: the insufficient quick adjustment ability of power supply, the unbalance in regulating the peak and off-peak period of the power grid; the immaturity in integrating various kinds of new clean energy into the grid.

(2) The construction of ultra-high voltage power grids is still in its infancy, with the planning, designing, construction, running, control and some other aspects of extra-high tension line still not developed into a standard system in technology. The synchronism of regional power grids has not been fully realized, the technical means for dynamic monitoring, evaluation, diagnosing, and assisting policy-making of large power network is still not effective, the flexible alternating current transmission technology waits for further application. All the above factors restrict the flexibility of power grid control method, and weaken the disaster response and restoring capability of power transmission network in some extent.

(3) The grid structure of the distribution network is relatively weak. This is mainly shown in the following: the power supplying ability of the distribution network in some cities is insufficient, the automatic coverage rate of the distribution network remains low, the electricity telecommunication equipment is backward, the losses of the distribution network is excessively high, the problem in power supply reliability and electrical energy quality remains severe. Moreover, the relevant technological and management system demands perfection urgently, the distribution network's bidirectional interactive support for the user is still insufficient and the research on the application of micro electrical network technology is still not mature.

(4) The load development is not balanced and the informationization degree differs greatly in different areas. The current electrical energy measurement mainly focuses on the electricity bill computation, resulting in an inadequate fineness for the collection of the user's measurement data as well as insufficient in-depth application of these data. The equipment and management level of the urban and rural grid and its directional interaction with the user is not sufficient enough to satisfy the demands of the modern power grid.

(5) Influenced by the industrialization and urbanization process as well as the improvement of people's living standards, the energy load will continue to grow, thus requiring the further transformation of city power network. Moreover, with the electricity-powered cars and energy-storing equipments being combined into the distribution network, the type and nature of load is also undergoing some change.

(6) Although the overall technical level of the transformer substation automation system is relatively advanced, its comprehensive utilization efficiency has not been fully realized. The pilot of the digitalized substation has been finished, but has not been widely promoted.

(7) The construction of advanced dispatch center has started, yet further research is still needed in the operation and control of interconnected UHV power grid, the real-time monitoring and pre-warning of the operating state, the

coordination of dispatching planning and the coordination control of the whole and regional power grid.

(8) The development of China's electric power communication is unbalanced. In the field of power line communications, we have realized on the whole the application of optical fiber in electric lead passage and the establishment of a data transmission network, and have successfully built a nationwide network of power integrated communications services, but the distribution network is still lack of reliable, economic and practical way of communication.

China has conducted relevant research in the construction and operation management of modern power grid construction and put it into practice, thus providing solid technical support and equipment guarantee for the construction of a unified, strong intelligent power grid. Pilot work on smart grid has already been started. Development on the basic platform of integrated smart grid scheduling-based technical support system has been completed; large customer load management and low voltage power carrier concentrated meter reading system has been taken into application in about 9 million households, research of Power Data Collection System has been in full swing; We have launched the construction of a high-level control center, a unified information platform as well as the pilot program of a user-oriented smart grid. In terms of large-scale renewable energy grid and distributed energy storage, we have carried out in-depth researches of certain key technologies such as the monitoring of PV power generation and grid-connection control, established the simulation platform for the integration of wind power into the power network, set up relevant criteria such as the technical regulation for integrating the wind power substation into the power system, and conducted basic research in frontier topics such as electrochemical energy storage.

Traditional power system includes four sections: generation, transmission, distribution and utilization. The future power system, however, will be intelligent, distributive, visual, controllable, automatic, clean and customer-centered. In the side of power generating, there will be more and more new energy source involved, such as solar power, wind power, energy storage device for large power grid. In the side of the user, there will be distributed energy, micro-network technology, etc. After the application of intelligent power meters, the users will become the supplier of electrical energy instead of just a pure user. Many large commercial users or industrial users will also become managers of the electrical energy.

These changes brings new concept for power systems in the future. Smart grid focuses on power delivery, asset management and customer experience. By using sensor technology and digital communication to manage massive grid data, it makes the power grid visible, controllable, and automatically integrated, thus enhancing reliability and reducing operating costs. At the same time, it can meet the requirement of new energy, distributed electricity power and power storage and reduce the negative impact on the environment.

Whether in the current situation or future development of the smart grid in our country, the integration of new energy and clean energy generators into the power generation system will exert deep influence on the current power supply structure.

The transmission system mainly covers such aspects as wide dynamic monitoring, real-time online alarming, fault diagnosis and analysis, etc. The distribution system mainly includes upgrading of the distribution network's voltage class, synchronous detection of the distribution network, visualization and load management system, fast simulation system and the optimal control and coordination systems; the user system mainly focuses on intelligent marketing system, smart metering system and user's optimization management system; the load system mostly includes the development of new energy storage devices, access of distributed power source, design of user-friendly electrical appliances and the research and development of new electric cars. The construction of intelligent substation is a combination of technologies such as intelligent network communication, information sharing and integration and intelligent real-time protection system; the intelligent dispatching center relies on the visualization technology, coordination and optimization technology of operation, intelligent pre-warning control and scheduling safety technology to realize the optimization, coordinated operation and safe scheduling of the power grid.

To conclude, the development of China's smart grid requires scientific planning of the development layout of the power grid. We should lay emphasis on both development of distributed power and the construction of large energy base, and combine the construction of ultra-high voltage transmission grid with the upgrading of distribution networks, the dynamic monitoring management of the whole network with the development of new energy storage devices, the intelligent transformer substation with the construction of integrated dispatching center and information collection system of power grid, so as to achieve the goal of secure, safe-healing, compatible, efficient and high-quality operation of grid in China.

(2) **Application of Smart Grid Technology in the Shanghai Expo**

It is known that the smart grid applied in the Shanghai Expo is the first demonstration project of its field in China. This smart grid project includes 9 subprojects, which include new energy access (East China Sea Bridge wind farm, Chongming solar PV, solar PV Expo venues), energy storage systems, intelligent substation, distribution automation, troubleshooting management system (TCM), power quality monitoring, electric information acquisition, intelligent buildings and intelligent home electric, electric vehicle charging and discharging and connecting into the grid. A number of records were set in this project, including 650 Ah sodium-sulfur battery with world's largest single capacity, China's first electric vehicle charging and discharging station, China's first set of TCM in the power industry. The East China Sea Bridge Offshore Wind Farm (100 MW) is Asia's first grid-connected offshore wind power project and Shanghai Chongming Island Vanguard Village PV power plant (1 MW) is the first PV feed-in demonstration project in China.

The Expo smart grid demonstration area included nine model projects and four demonstration projects. The nine model projects were contained in the smart grid of the Expo and four demonstration projects were open to visitors who could to learn the smart grid closely and visually at multi-angle. Especially in the basement of the

Fig. 2.99 State Grid Pavilion in the Expo

State Grid Pavilion (Fig. 2.99), through the large glass windows, visitors can see how a 110 kV substation operates. As the first Intelligent Substation served for the Expo, known as the “heart of energy” of the Expo Puxi site, all processes such as information collection, transmission, processing and output adopted digital and intelligent workflow after it was put into operation.

Four large PV power stations in the Expo Park corresponding to China Pavilion, Theme Pavilion, Expo Center Pavilion and the Pavilion of Future were connected to the grid, with the total capacity of 4.687 MW, of which the Theme Pavilion had a PV generating capacity of 2,825 kW, becoming the largest single BIPV power station in Asia. During the World Expo, the generating capacity in Shanghai from grid-connected clean energy reached 146.4 MW, including 139.4 MW by wind power, 7 MW by PV power, which can provide one-third of electricity consumed in the Expo site.

For the demonstrated electric vehicle charging and discharging stations and the interactive system between electric vehicles and the grid in the Expo area, the controlled two-way flow of energy can be achieved between the electric vehicles and the grid. Dynamic response can be obtained based on the grid running condition and the control command, which displayed the application potential of electric vehicles as distributed mobile energy storage.

The significance of constructing a smart grid demonstration area in the Expo site is to promote the concept of energy-saving and facilitate the application of relevant high-techs. The smart grid technology, with ultra-high voltage grids as its main frame, greatly helped the intensive development, long-distance transmission and highly efficient utilization of those clean energies like wind power and solar power, so as to achieve the standardization of distributive powers, power storage devices and the connection of power devices to grids as well as to achieve the intelligence of grid operating controls. At the same time, it greatly reduced the power loss in the long-distance transmission and therefore occupied a commanding height of the new economic growth.

It's a good start for smart grid in the Expo to withstand the test of high temperature, thunder and lightning. It's expected that by 2020, in a strong smart grid scenario, the power loss of the whole society can be reduced by 18.95 billion Yuan.

(3) **Suggestions on Smart Grid Technology**

According to statistics of China Electricity Council, in 2008, China's total investment in electricity infrastructure was 576.329 billion Yuan, an increase of 1.52 % year on year, among which investment in power source was 287.873 billion Yuan, down by 10.78 % year on year, the amount of power grid investment reached 288.456 billion Yuan, an increase of 17.69 % yoy. The proportion of the power grid infrastructure investment reached 50 % of the total electric investment for the first time. At the same time, with the help of 4 trillion investment plan in 2009, the construction of power grid will be the focus of investment in the future, with the proportion of grid investment expected to increase by 10 %. But the proportion of smart grid is also rather small.

The National Grid recently announced for the first time the phase targets of smart grid, as follows:

① 2009–2010—planning and pilot phase. During this period, focus on establishing smart grid development plan, formulating technical and management standards, conducting research and development of key technologies and equipment as well as pilots in various stages.
② 2011–2015—overall construction phase. During this period, the construction of ultra-high voltage power grid and urban-rural distribution network will be speed up to achieve the initial formation of the smart grid system operation control and interactive services, and to make major breakthroughs and wide applications of key technology and equipment.
③ 2016–2020—lifting phase. The aim at this stage is to establish a unified fully integrated smart grid, ensuring that the resource allocation capacity, safety level and operating efficiency of the grid and the interactivity between the grid, power supply and users will increase significantly, with technology and equipment reaching the international advanced level.

2.2.3.3 Non-electric Air-Conditioning Technology

(1) **Introduction to Non-electric Air-Conditioning**

Non-electric air-conditioning is to achieve refrigeration directly with thermal energy. Traditionally, a cooling process of an electric air-conditioning has to go through a series of energy conversion process from thermal to mechanical energy, then to electric energy, and finally back to mechanical energy before the cooling is made. The five energy conversions involved in this process inevitably results in energy loss, and the amount of carbon dioxide emissions from coal-powered

Fig. 2.100 Non-electric air conditioning unit in Broad Pavilion

Fig. 2.101 Broad non-electric air conditioning unit used in the Expo site

air-conditioning are five times as large as that of the non-electric ones. Therefore, non-electric air-conditioning can not only save primary energy, but also effectively reduce the carbon dioxide emissions by four times. To sum up, the feature of a non-electric air-conditioning is it is powered by natural gas, waste-heat power, or solar energy instead of electricity and that makes it the cleanest air-conditioning.

(2) **Applications of Non-electric Air-Conditioning in the Expo**

It is reported that Broad Non-electric Air-conditioning has been applied in all the pavilions in the Shanghai Expo (Figs. 2.100 and 2.101). Besides, the world's leading energy-saving mode of air-conditioning was introduced. 22 energy centers was built in the Expo site, with 14 in Pudong and 8 in Puxi, distributed in each area to provide air-conditioning service for over 200 pavilions, resulted in the reduction of 73,000 tons of CO_2 emissions, equal to the amount absorbed by 4 million trees in half a year, or the forest coverage of 8 Shanghai World Expo sites. By doing this, the Shanghai Expo met its goal of being a "Low-Carbon Expo" in a real sense.

(3) **Promotion of non-electric air-conditioning technology**

In addition to the difference that non-electric air-conditioning uses natural gas as the energy source, a non-electric air-conditioning uses lithium bromide as the refrigerant instead of Freon. Lithium bromide is a kind of saline material extracted from seawater, with good water absorption. When heated, lithium bromide solution will release vapor, which will be condensed into water, which is sprayed over the copper tubes for cooling. This is how cooling is made. Water takes away the heat of the air conditioning system and becomes vapor again, which is absorbed by lithium bromide solution, and the diluted lithium bromide solution is heated and water vapor is regenerated to repeat this cooling cycle. Non-electric air-conditioning is not completely free from the use of electricity, but its consumption of electricity is only 10 % as much as that of the common electric air-conditioning.

For most of us, non-electric air-conditioning is new term. At present, the majority of people are taking a wait-and-see attitude in deciding whether to give up the traditional air-conditioning and use a non-electric one. The reasons for this hesitation are the traditional air-conditioning is more convenient and technically mature. Anyway, people need time to accept new things. Another reason is the concern for the costs.

The key factor in popularizing the non-electric air-conditioning into thousands of households lies in the change of people's concepts. In spite of its higher cost than traditional air-conditioning, non-electric air-conditioning will save much more electricity from the perspective of the whole society. In addition to the electricity bills it saves, the application of non-electric air-conditioning will reduce carbon emissions, and thus reduce the potential costs of the nation's environment management and dealing with the side effects caused by environmental deterioration. However, it takes time for people to foster a low-carbon concept of environmental protection. Therefore, there will definitely be a long way ahead before every one can abandon his or her old lifestyle and embrace the new lifestyle of low-carbon.

2.3 Low-Carbon Transportation

2.3.1 Low-Carbon in Transportation

2.3.1.1 Introduction to Low-Carbon Transportation

The low-carbon transportation includes three parts: control of the pavement noise, increase of the driving skill and the road material improvement.

In terms of driving technique, one must have good driving habits, make slow acceleration when starting, maintaining a smooth driving on the road, with stable and smooth movements. And a correct judgment on the traffics on the road is necessary. Second, pay attention to maintain a good condition of your car, including warming up the vehicle properly before driving it, accelerate after stable driving

1–2 km and regular cleaning air filter, ensure that the vehicle is in good condition and therefore be fuel-efficient.

Try to use public transport as much as possible. More walking, bicycle riding or light railway and subway, and less drive.

Low-carbon road involves three aspects: the first is to improve the quality of road construction and road management so as to extend the service life of the road; Second, to develop low-carbon construction and the road surface materials; third, to recycle the waste road materials.

2.3.1.2 Application of Low-Carbon Transport in the Expo

The application of low-carbon transport in the Expo is mainly reflected in the road material. 60 % of the city demolition wastes are used as the road materials, which turned the waste into things of value while putting low-carbon concept into practice.

2.3.2 New Energy Vehicles

2.3.2.1 Introduction

The *Management Rules for New Energy Automobile Manufacturing Companies and Product Access* (the Rules) was carried out officially on July 1, 2009. The Rules emphasizes that new energy vehicles use unconventional fuels as a power source (or use conventional fuels but apply new car power plant). By integrating the advanced technology in vehicle's power control and drive, these cars are made with advanced technical principles featuring new technology and new structure. New energy vehicles include Alternative Fuel vehicles (AFV), Hybrid electric vehicles (HEV), pure electric vehicles (PEV) and solar vehicles. AFV further consist of Methanol Fuel vehicles (MFV), Compressed Natural Gas Fuel vehicles (CNGV) and hydrogen engine vehicles.

With the increase of the vehicle population in China, the proportion of demand for petrol and diesel consumed by vehicles is rising. It is estimated that, by 2020, the petrol needed by vehicles will account for over 50 % of the total oil consumption. Therefore, in order to ease the pressure in oil import and environmental protection, it is of great significance for us to advance the research and promotion of fuel cell vehicles.

(1) **Alternative Fuel Vehicles**

At present, there are mainly two types of alternative fuel cell vehicles. One is direct fuel cell vehicle (DFCV). In the cell of DFCV, the fuel, which could be hydrogen or methanol, will be oxidized directly by electrochemistry. Another is called processed fuel cell vehicle (PFCV). In the system of PFCV, the fuel, which could be gasoline, methanol or propane, will be first converted to hydrogen by means of a fuel processor. Then the hydrogen will be conveyed to the cell to generate electricity. Currently what has attracted more global concern is the

PEMFCs that directly use hydrogen as its fuel, and there are more successful examples both in the lab and in application.

Methanol vehicles use methanol fuel as the power instead of petroleum fuel. This kind of vehicles has been applied in other countries for years and very mature in its control system technology. In recent years, due to the shortage of petroleum resource and the intensified diversification of automobile fuel, methanol vehicles again have been listed as one of the major alternatives.

Compressed Natural Gas (CNG) vehicles use CNG, Liquefied Petroleum Gas (LPG) and Liquefied Natural Gas (LNG) as the fuel. For the past few years, governments of different countries all have actively been seeking to solve this problem and started to adjust the automobile fuel structure. Because of its good performance in discharge, ability to adjust automobile fuel structure, low operating cost, mature technology and stability, the CNG vehicles are universally acknowledged as the optimal AFV.

Hydrogen Fuel Cell Vehicles use hydrogen as the fuel; in fuel-cell conversion, the hydrogen is turned into electricity through chemical reaction which then powers electric motors. The energy of the cell is generated via the chemical action between hydrogen and oxide instead of combustion and directly converts to electric energy. This process produces no hazardous substance thus hydrogen fuel cell vehicles are pollution-free automobiles. Since the energy conversion efficiency of hydrogen fuel cell is 2–3 times higher than internal-combustion engine, hydrogen fuel cell vehicles are the ideal ones on the aspect of energy utilization and environmental protection. At present, the most attractive application orientation for hydrogen fuel cell is to be the power supply for low-discharge and diversified electric automobiles. Large automobile manufacturers like Toyota, Honda, General Motors and Daimler Chrysler have successively made tremendous investments in developing hydrogen fuel cell vehicles. Take General Motors for instance, it has made the investment of 1 billion US dollars in a research and develop project of Cell Electric Automobile. Although the manufacturers are enthusiastically hoping that hydrogen fuel cell vehicles will enter the market within a few years, none of them actually reaches the stage of commercialization. Constrained by various conditions, problems in the hydrogen fuel cell vehicles are as follows:

- As the main fuel for the cells, Hydrogen (H_2) does not exist solely in the natural world and difficult to be transported. Therefore, it is widely acknowledged that the preparation and transportation of hydrogen can be the biggest problem in promoting this technology.
- Currently, the hydrogen cells have to be catalyzed by a large amount of platinum, which is a kind of the precious metal and is limited in its reserves. Based on the researches at the moment, the amount of platinum utilization cannot drop in a short period. This is also a technical problem that we have encountered.
- The manufacturing cost for hydrogen fuel cell is relatively higher so that ordinary clients may hardly afford it. Take the unit power cost of the engine for example, the current cost of fuel cells is about US $500 per kW, while the cost

of internal combustion engine is only US $11–27 per kW. The cost of utilization at present is unbearable for most people thus becoming another problem.

- Suppose that fuel cells will replace the internal combustion engines in the future, all the combustion engine plants will have to be closed. This indicates that promoting fuel cell vehicles on a large scale requires an adjustment and a new overall arrangement in a wide scope. The replacement could affect not only gas station worldwide but also many industries such as petrochemical industry, power industry, machinery industry and electronics industry. It will negatively affect those industries while create some new opportunities and finally generate a huge impact on the world political and economic structure.

(2) **Hybrid-Electric Vehicles**

HEV refers to vehicles that consume traditional fuels and are supported by motors/engines to improve low rates of power output and fuel consumption. Based on different fuel types, HEVs can be divided into gasoline hybrid-vehicles and diesel hybrid ones. At present, the former dominates the domestic market, while the latter are also developing fast in the international market. Figure 2.102 shows the working mechanism of a tandem hybrid electric vehicle.

(i) Historical development

Currently, the commonly used fuel-driven motor vehicles have many demerits. The statistics show that on more than 80 % of roads, an ordinary car only uses 40 % of its power capacity. In urban areas, the figure will even fall to 25 %. More seriously, its emissions will pose threats to the environment. Since the 1990s, the call of world for environmental protection has been growing louder and a variety of electric vehicles have been emerging. Although it is generally believed that electric vehicles will dominate the future vehicle market, problems with the current battery technology have hindered the use and spread of electric vehicles. As the power density of battery is about a hundred times less than that of gasoline, which is far from the required value, experts estimate that within 10 years electric vehicles will not replace fuel electric ones (unless there is a major breakthrough in non-fuel battery technology).

Engineers have been forced to develop a vehicle called HEV, which combines merits of both hybrid and electric vehicles. The so-called hybrid-electric device refers to the combined application of a motor and an APU to one car as the driving force; the APU is a small fuel engine or a set of electric generators. To put it simple and vivid, engineers try to make the conventional engine smaller so that part of the power can be generated by the battery-motor system. This device not only gives a full play to the traditional engine that can work continuously for long hours and function well in terms of power generation, but also has the advantages of the electric motor that is pollution-free and producing little noise. Both are mutually collaborative and supportive, and therefore the thermal efficiency of the vehicle can increase by more than 10 % and its emissions can be reduced by more than 30 %.

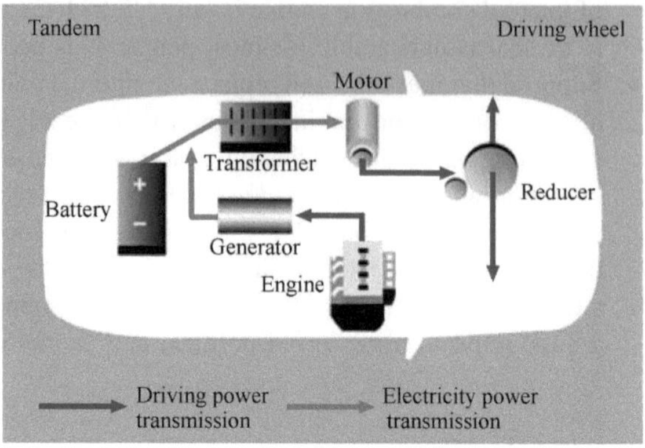

Fig. 2.102 Working principle of hybrid electric vehicle

(ii) Advantages

a. With the hybrid-electric power, we can determine the maximum power of the combustion engine according to the power acquired in daily use. At this point, the engine is working with low fuel consumption and less pollution. If the power generated by the high-power internal combustion engine is less than required, the battery can provide the rest. When the load is low, the surplus power can generate electricity to charge the battery. The internal combustion engine can work continuously and the battery can be charged simultaneously, so the HEVs can travel as far as the ordinary ones.
b. With the battery, it can be very easy to recollect the energy generated in the operations of braking, going to the downhill and idling.
c. In the downtown areas, we can shut down the internal combustion engine, so the vehicle will only be driven by the battery so as to achieve "zero" emissions.
d. With the internal combustion engine, it can be very convenient to solve problems encountered by PEV, such as the energy-consuming air-conditioning, heating, defrosting, etc.
e. The vehicle can be refueled in existing gas stations on which there is no need to invest more.
f. The battery can work in good condition without such problems as over-charge and over-discharge, so its life can be prolonged and costs reduced.

(iii) Disadvantages

Basically, it can not save fuel in long-distance travel with a high speed. That is to say, its energy saving effect is not obvious.

(3) **Pure electric vehicles**

Electric vehicles are those mainly driven by electric power (super capacitor vehicles belong to this category as well). Most vehicles are directly driven by electric engines; with some, electric engines are installed in the engine compartment; there are still some that have their wheels as the rotor of the four engines.

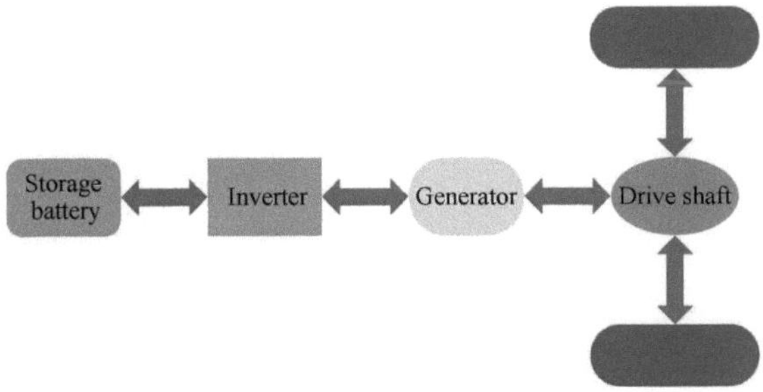

Fig. 2.103 Schematic diagram of a pure electric vehicle

The difficulty lies in the electricity storage technology. The fact that the electric power can be obtained from a variety of primary energy resources such as coal, nuclear, water, wind, light, heat, etc. can lift people out of an increasing concern over the decreasing oil resources. Figure 2.103 shows the schematic diagram of a pure electric vehicle.

Advantages: Electric vehicles can make the best use of surplus electric power in the off-peak hours in the evening, so that the power generation equipment can be used throughout the whole day, thus bringing great economic benefits. Research shows that when the crude oil undergoes the process of being refined and sent to the power plants for generating electricity which is then charged into the battery to drive the battery-powered vehicles, it has a higher energy utility efficiency than the same crude oil that is refined into gasoline which is consumed in the engines to drive the vehicles. Therefore, it helps greatly save energy and reduce carbon dioxide emissions.

Disadvantages: At present the energy stored per unit weight of the storage battery is fairly low, due to the high price of batteries for electric vehicles and no economic scale, the vehicles are expensive. As to the cost to use, some trial results show that electric vehicles cost more than ordinary ones, while some cost only 1/3 of the ordinary vehicles, which depends on the service life of batteries and prices of local oil and electricity.

(4) **Solar vehicles**

Solar vehicles are a type of vehicle powered by solar energy (Fig. 2.104). Compared with the traditional engine-driven vehicles, solar vehicles have truly achieved zero emissions. Due to its characteristics of environmental friendly, it has been advocated and promoted in many countries around the world, leading to the booming of the solar vehicle industry. With the solar energy vehicles, photoelectricity replaces oil gasoline to provide the needed power so as to save the limited oil resources. In the daytime, the solar batteries transform the solar energy into electricity that is automatically stored in the power batteries. At night, the batteries can be charged making use of the off-peak hours (220 V). So far, solar energy technology applied in vehicles includes two aspects, one is as the driving force and the other is used as the source of energy for the vehicle supplementary devices.

Fig. 2.104 Solar vehicle

The advantages of the solar vehicle are that it produces no noise or pollution. As solar vehicle does not consume gasoline, no harmful gas is discharged into the atmosphere. Also because it does not install internal combustion engine, no rumbles from the working engine of a traditional vehicle will be heard. A solar vehicle is simple in structure with no such parts and components as internal combustion engine, clutch, mechanical transmission, transmission shaft, radiator and exhaust pipe. As a result, it is easier to manufacture than traditional vehicles.

2.3.2.2 Application of New Energy Vehicles in the Shanghai Expo

Numerous clean energy bus lines were available in and around the Expo site, which corresponded with the goal of "zero emission inside the Expo site and low emission in the neighboring area." The hydrogen refueling station replaced the traditional gas station to serve the clean energy vehicles. The number of such vehicles actually applied amounted over 1,000, including ultra capacitor vehicles, fuel cell vehicles, PEV, hybrid vehicles, and solar vehicles, etc.

(1) **Applications in the Expo Site**

Figure 2.105 illustrates the amount of different new energy vehicles applied in the Expo Site.

Various kinds of new energy vehicles serving as the shuttles took visitors to any part in the Expo site and across the Huangpu River. As one of the first experimental cities in promoting and demonstrating the energy-saving and new energy vehicles, Shanghai has cooperated with the National Science and Technology Department to carry out the demonstrational operation involving 1,017 various kinds of new energy vehicles during the Expo.

New energy vehicles applied in the Expo site are classified as: sightseeing, garbage collection, patrol, public transport, VIP reception, etc. See below the pictures of different types of vehicles in Figs. 2.106, 2.107, 2.108, 2.109, 2.110, and 2.111.

The electric vehicles are also equipped with special charging station (pillar). The most common ones seen in the Expo Site are the charging stations for electric

Fig. 2.105 New energy vehicles in the Expo by classified statistics

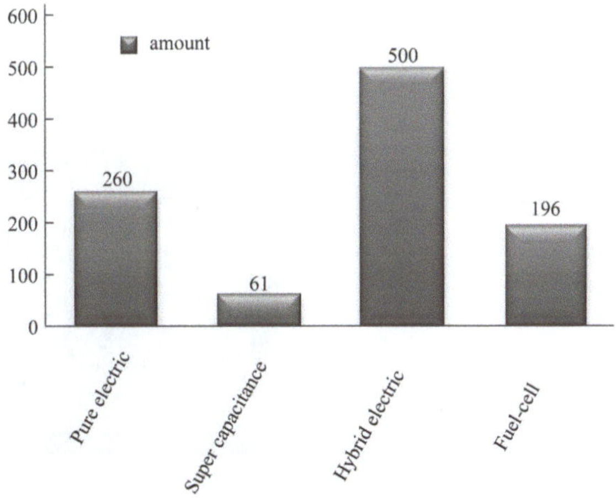

Fig. 2.106 Ultra-capacitor bus in the Expo site

Fig. 2.107 Fuel cell sight-seeing cars

Fig. 2.108 VIP reception car with hybrid electric system

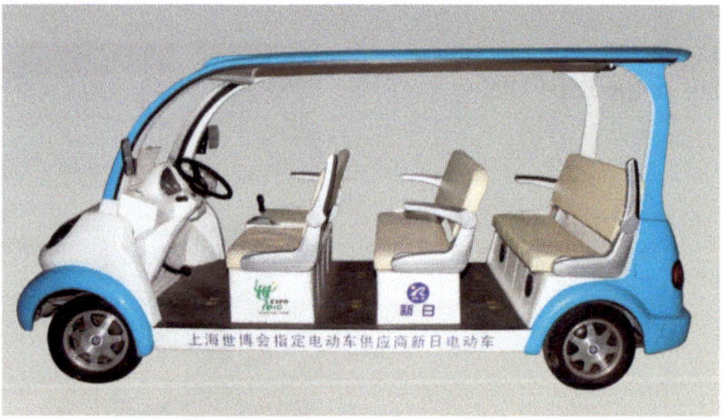

Fig. 2.109 Pure electric car between pavilions

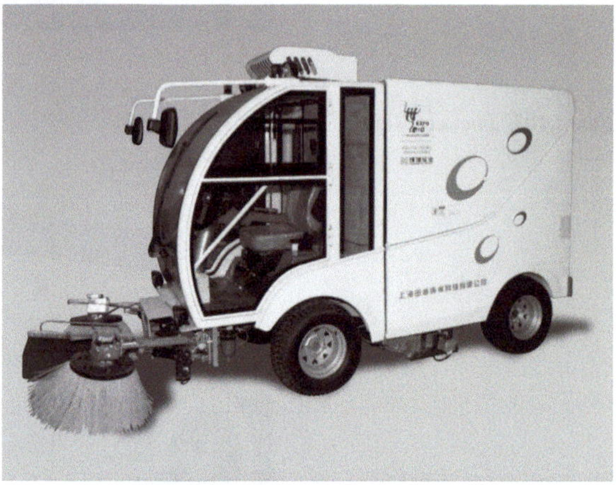

Fig. 2.110 Cleaning truck with full-electric system

cleaning trucks (Fig. 2.112) and 5 kW AC charging stations for electric cars (Fig. 2.113). Such charging stations have an output voltage of 220 V AC, with a maximum current of 22 A. It has the following features:

Fig. 2.111 Four electric patrol car in the Expo site

Fig. 2.112 Charging station for electric cleaning truck

1. It is easy to operate; 2. It is flexible in choosing the charging method; 3. It can cater to the need of commercial operation, with self-lock device and emergency stop button, the charge can be finished unattended; 4. It can ensure charging security due to its function of electrical connection confirmation with the vehicles; 5. It can indicate information about voltage and electric current and mode during charging.

(2) **Demonstrations in the pavilions**

The information of different types of new energy vehicles displayed in the pavilions is shown in Table 2.6.

Figure 2.114 shows Roewe E1 electric mini-car displayed in China Pavilion. Figure 2.115 shows Smartfortwo electric car exhibited in Switzerland Pavilion.

Fig. 2.113 5 kW AC charging station for electric vehicles

Table 2.6 Different types of new energy vehicles displayed in the pavilions

Pavilions	Types	Features
Switzerland Pavilion, Pavilion of Urban Planet, Austria Pavilion	Smartfortwo Electric cars	Electro-motor of 40 hp,a max speed of 100 km/h
Japan Pavilion	Pure electric vehicles	Pure electric-driven, Toyota RAV4, estimated to be launched on the market in 2012
Portugal Pavilion		METROBuddy mini pure electric vehicle, continuous travel distance is120 km
France Pavilion		This pint-sized car is only 2.5 m long, which can hold four people
Sweden Pavilion	Volvo hybrid vehicles	Parallel hybrid technology
Monaco Pavilion	VenturiVolage latest high-end concept car	Electric vehicle, Quattro technology plus active suspension technology
China Pavilion	Roewe E1 mini electric car	

Figure 2.116 shows Toyota pure electric car displayed in Japan Pavilion. Figure 2.117 shows Buddy mini pure electric car in Portugal Pavilion. Figure 2.118 shows Volvo Hybrid electric City bus displayed in Sweden Pavilion.

Fig. 2.114 Roewe E1 electric mini-car displayed in China Pavilion

Fig. 2.115 Smartfortwo electric car exhibited in Switzerland Pavilion

2.3.2.3 Review and Analysis of New Energy Vehicles Application in the Expo

The quick charging device installed on the top of the super capacitor bus in the Expo site takes advantage of the interval when passengers get on and off the bus to complete its charge. According to the vehicle maintenance engineer, it only takes about 2 min to finish the charge which can sustain its driving for 3–5 km.

At the beginning of the Expo, there were altogether 36 super capacitor buses, 120 electric cars and 70 fuel cell sightseeing cars put into use in the Expo site for

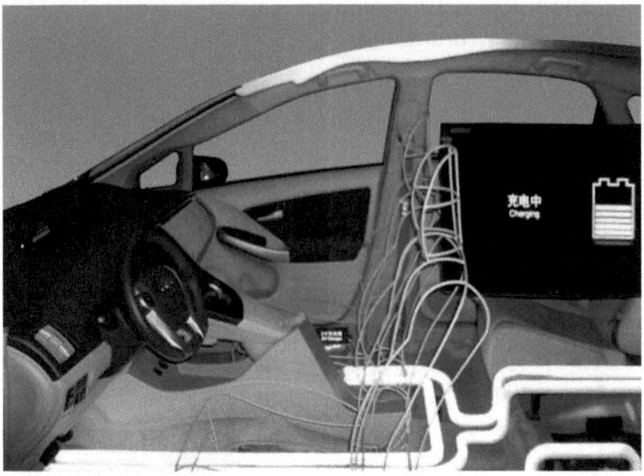

Fig. 2.116 Toyota pure electric car displayed in Japan Pavilion

Fig. 2.117 Buddy mini pure electric car in Portugal Pavilion

Fig. 2.118 Volvo hybrid electric City bus displayed in Sweden Pavilion

trial operation. However, as the number of passenger increased, more cars were needed. The organizer was quick in responding to this issue, on one hand, they continued to put more new energy automobiles into use, on the other hand, they brought together over 10 suppliers specializing in the research and development of pure electric automobiles to solve a lot of technical problems in a joint effort. As hot season arrived, nearly a thousand of such new-energy automobiles have passed the preliminary test posed by hot days, with a serviceability rate of 99.73 %, and work ratio of 99.59 %.

With thousands of new energy vehicles running in the Expo site, it is estimated to save about 10,000 tons of traditional fuel during the 6-month Expo. Besides, they also helped reduce the emissions of harmful substance by about 118 tons and greenhouse gases (mainly carbon dioxide) by about 28,400 tons. Among all these different types of new energy vehicles, the fuel cell car is completely free from the use of fuel, and is regarded as the trend of sustainable development of auto industry in the future and one of the most ideal solutions for the problems of global energy and global warming due to its zero emission, high efficiency and extensive fuel sources.

2.3.2.4 Current Status of New-Energy Vehicles

The key point of the new energy vehicles lies in its battery. According to a report, nickel-hydrogen batteries have taken up a 99 % market share among all the existing types of hybrid batteries, and the most commercialized is Toyota Prius. At present, the main manufacturers of automobile batteries around the world are PEVE and Sanyo from Japan. PEVE accounts for 85 % of the nickel-hydrogen batteries used in Hybrid automobiles. Currently, the main commercialized automobiles such as Prius, Alphard, Estima of Toyota, and Civic, and Insight of Honda have all adopted the PEVE nickel-hydrogen battery pack.

Speaking of the development of new energy automobiles, the Nickel-hydrogen battery is considered the most mature technology. During the next 3 years, new energy automobiles will remain as the mainstream, and after that the market will be divided among batteries of nickel-hydrogen, lithium-ion and hydrogen fuel. And in 5 years, nickel-hydrogen batteries will gradually be replaced by fuel cells. Nowadays, more and more auto manufacturers choose lithium battery for the new energy automobile. Lithium cell has the advantages of small volume, light weight, high working voltage, long life circle, zero pollution, and security, etc. Many world's well-known auto manufacturers such as Ford, Kreisler, Toyota, Hyundai, Courreges, Ventury, etc., are committed to the development of new energy automobiles using lithium cell. Meanwhile, some domestic car manufacturers also aim to develop hybrid automobiles and pure electric automobiles coupled with lithium cell batteries.

2.4 Technologies in Water Treatment

2.4.1 Water Purification

2.4.1.1 Overview of Water Purification

As most of the water bodies in the world are seriously polluted, human beings are facing increasingly shortage of water resource. Conventional water treatment method can not ensure the high quality drinking water. Moreover, secondary pollution exists in the municipal water supply, such as water tank in the high-rise, long line of pipes for tap water may cause the potential rust, scale and micro contamination. To meet the demand for clean water, various brands of water purifiers have thus been invented.

Based on the water treatment method, the purifiers are involved in the following 11 methods:

(1) **Softening**: is to remove or reduce the hardness (mainly calcium and magnesium ions) of the water. Soften only makes water soft but can not improve water quality.

(2) **Distillation:** it refers to boiling the water to produce water vapor. The vapor contacts a cool surface where it condenses as a liquid. Distilled water is very safe drinking water. Because distilled water contains no minerals, those who criticize it point that this may speed up the process of aging. Besides, distillation method has the disadvantages of high cost, high energy consumption and hard to remove the volatile substance in the water.

(3) **Boiling**: an ancient purification method and widely used domestically. Water is heated hot enough and long enough to kill micro-organisms that normally live in water at room temperature. But boiling does not remove chemical substances or heavy metals in the water even the content is very low. Therefore, it's not safe for drinking.

(4) **Magnetization:** treat water taking advantage of magnetic field effect. So far, magnetization treatment of water in China has been still in the initial stage of practice and research. There is no requirement for magnetization in the purifiers in other countries as this is not in the scope of purification but medical treatment.

(5) **Mineralization**: is to add the beneficial mineral elements (e.g., calcium, zinc, etc.) to the purified water with the purpose of strengthen the health benefit of mineral water. The purifiers on the market normally add medical stone into the water to achieve the mineralization. But this is still controversial as for the effect.

(6) **Ozone and ultraviolet disinfection**: They only kill the organisms but can not remove the heavy metals and chemicals in the water. The killed bacterium remains are still left in the water to become the pyrogen.

(7) **Electrolysis and Activation**: This is a new invention in Japan. Water is purified first and then electrolyzed and activated. After this process, the water

corresponds with the PH inside the body of human beings and therefore has a health function. For the purifier, the effect requires further research.

(8) **Activated carbon absorption**: including three forms:

- Granular activated carbon: commonly used
- Silver activated carbon combination: with the effect of both absorption of organic contaminants and sterilization.
- Fiber activated carbon: with great microscopic bore structure, huge specific surface area and numerous functional groups.

(9) **Reverse osmosis membrane**: a membrane separation technology. Mechanical pressure is applied to an impure solution to force pure water through a semi-permeable membrane. While pollutants are removed. Reverse osmosis is theoretically the most thorough method of large scale water purification available, although perfect semi-permeable membranes are difficult to create. This system cost much and requires much routine service, such as monitoring and membrane change.

(10) **Microfiltration and ultrafiltration:** Microfiltration uses cellulose or high polymer material as the microfiltration membrane to hold back the particles, organisms and colloid in the water by using its uniform microscopic bore. Both microfiltration and ultrafiltration are membrane separation technology, with no distinct limit.

(11) **Compound technology**

When one technique can hardly remove the harmful substances from the water, two or more techniques are adopted, this is called compound technology, which involves some of the above technologies. The purifiers with membrane composite technology proves to have good performance, which is very efficient in filtering out the micro organisms. Some of the quality ones are very popular among the consumers.

From the above classification of the water purifiers it is easy to conclude that household water purifier is actually a miniaturized deep water treatment device. The process consists of pre-filtration (coarse filtration), absorption, fine filtration (microfiltration, ultra-filtration and reverse osmosis), of which absorption and fine filtration are the major means of removing the organic matter and odor and color in water. Household purifier aims at millions of households, but the current situation is, the regional water quality and water pressure vary widely, and users usually lack the necessary operation knowledge, plus the current product design can not cope with the ever-changing situation. So the water purifier manufacturers should take into consideration how to educate users with certain knowledge, proper installation and operation of water purifier, so as to achieve the desired water quality, and thus reassure the users.

Fig. 2.119 Water treatment device in Japan Pavilion

2.4.1.2 Application of Water Purification in the Shanghai Expo

Japan Pavilion exhibited a water treatment device known as MBR technology shown in Fig. 2.119, which is one of the most advanced water treatment technologies in the world. Unlike before when the reclaimed water from sewage could only be used for toilet flushing, plant watering and car-washing, this technology is upgraded to provide drinking water. Inside the Osaka Pavilion of the UBPA, tourists have an interesting experience of converting sewage into purified water by just pedaling a specially designed bicycle.

Faced with the water shortage as Japan is, Singapore has also made great efforts in water treatment. In Singapore, there are 15 reservoirs and one water-collecting system that can defend flood in case of rainstorm. With the completion of Punggol reservoir and Serangoon reservoir in 2011, the catchment area, currently accounting for half of Singapore's land, will be enlarged to take up 2/3 of the whole country's area. Now they are promoting a technology which can turn the wastewater into drinking water directly. Moreover, there is also a widespread water-saving proposal in Singapore to convince people to use water in a rational way. At present, the per capita daily consumption of domestic water in Singapore has been cut down from 165 L in 2003 to 155 L, and with an aim to lower the amount to 140 by the year of 2030.

In the happy street of Netherlands Pavilion, the world's smallest water purification system is displayed (Fig. 2.120). It is about 1 m wide and 2 m high. Though the activated carbon technology, it can produce the highest-level drinking water, and serve the tourists with its purified water, the source of which is from Huangpu River.

Norway Pavilion exhibited a solar-powered water purification device (Fig. 2.121), which converts the contaminated rain water into the safe purified drinking water. This device can remove all the contaminants in the water, including bacteria, viruses, parasites and so on. In the purification process, no environmentally hazardous substance is used. The system can be loaded into the standard-sized containers for transport.

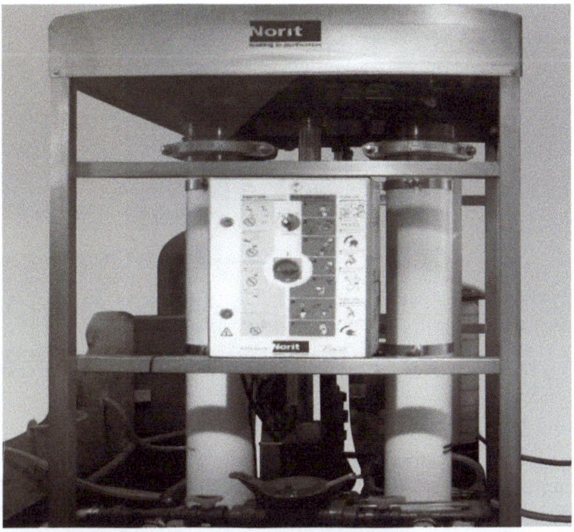

Fig. 2.120 Water treatment device in Netherlands Pavilion

Fig. 2.121 Water treatment device in Norway Pavilion

At present, the drinking water supplied in the Expo site is the direct drinking water system independently developed by our own country, which uses the "Activated Carbon + PVC (polyvinyl chloride) alloy ultrafiltration membrane + UV" combination treatment technology. Among the multiprocessing procedures and over 200 criteria, membrane technology has the highest technical content. Besides, the recycling project for domestic water and rainwater in Pavilion of Future, with a scale of 120 tons per day, adopts the world's advanced alloy ultrafiltration membrane bioreactor. The device has a very high performance and price ratio, with an annual water treatment capacity of over 40,000 tons. All the reclaimed water after treatment is used for toilet flushing and plants watering. Figure 2.122 shows the water purification device in the Expo. Figure 2.123 shows the cup cleaner water-saving toilet in United Nations Pavilion.

Fig. 2.122 Drinking water purification device in the Expo site

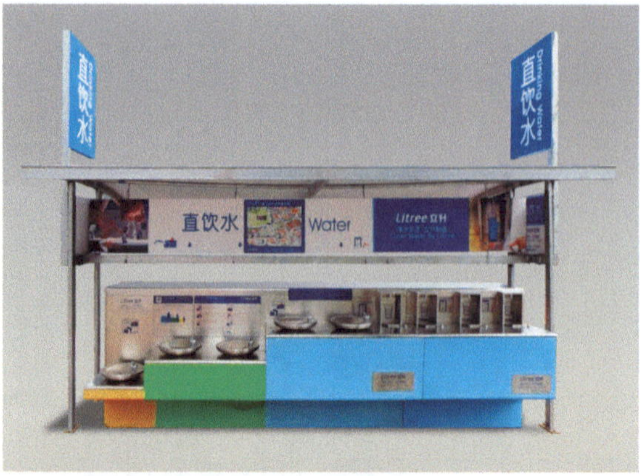

Fig. 2.123 The cup cleaner water-saving toilet in United Nations Pavilion

Water saving and recycling is another theme of the World Expo. The sanitary ware and irrigation equipment are all water-saving devices. The permeable floor in the Expo site can effectively reduce the runoff and pollution efflux to the pipe network of urban rainwater. The permanent pavilions such as Expo Center, Performance Center, Theme Pavilion and China Pavilion, together with the Expo Axis have all resorted to a roof rainwater utilization system, by which the rainwater and sewage collection and processing rate can reach a full 100 %, with an estimated water-saving amount of over 1 million m^3. Figure 2.124 shows water treatment application in the Expo site (statistics by continent).

Fig. 2.124 Water treatment application in the Expo site (statistics by continent)

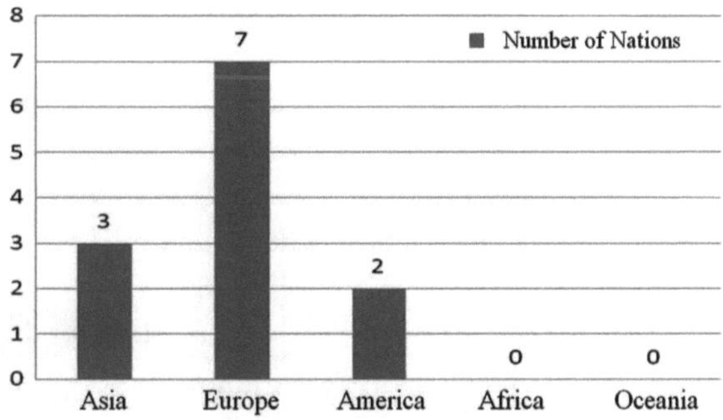

2.4.2 Rainwater Treatment

2.4.2.1 Brief Introduction

Rainwater treatment is the purification process for Surface runoff water in the urban area (rainwater, snow water and irrigating water). Based on the different pipe system drainage system (separate or confluent system), rainwater treatment can be processed independently or combined with the domestic sewage and industrial wastes. When treated independently, the rainwater catchment and sedimentation pond is needed, with grilling in front and filter pond in the back. Water flow rate in the pool should be less than 0.01 m/s, for a period of not less than 1–2 h. After treatment, the biological oxygen demand (BOD) can be reduced by 50–70 %. In the latter case (combined treatment), the water amount in the grilling and sand setting pond is calculated based on the total amount of water. The residence time in the sedimentation pond shall not be less than 1 h. If the rainwater after treatment is used for as the industrial circulating water, the processing requirement should be calculated by the circulating water. The discharge port can be placed on the water bank as needed.

2.4.2.2 Rainwater Treatment Application in the Shanghai Expo

(1) Collection and Application of Roof Rainwater

Located in the core area of the Expo, the four permanent pavilions like Expo Center, Cultural Center, the Theme Pavilion and the China National Pavilion, together with the Expo Axis and Shanghai Corporate Pavilion, have all engaged in the collection and application of roof rainwater.

The roof rainwater from the Expo Center will be reclaimed for road cleaning and plants and lawns watering (Fig. 2.125).

As for the rainwater harvesting system in the Expo Cultural Center (Fig. 2.126), it collects and further process the coagulated water from air-conditioning and roof rainwater, so that they can be used for road cleaning and plants irrigation.

Fig. 2.125 Expo center

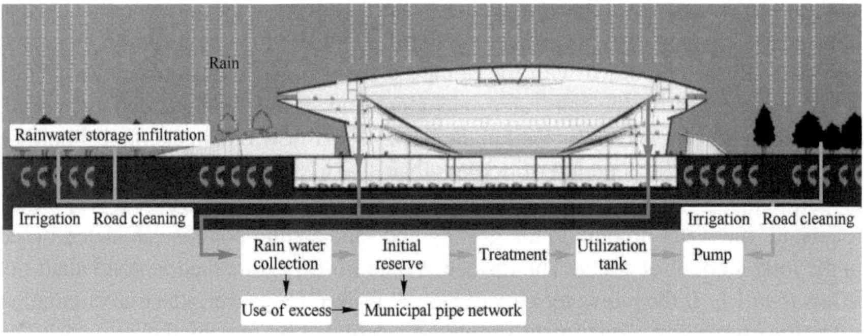

Fig. 2.126 Rainwater collection and utilization system in Expo Culture Center

By adopting the programmed control of spray and drip irrigation, the utility rate of water resources can be significantly improved.

Large amount of rainwater from the Expo Axis is stored in its basement. After being filtered, the rainwater can not only be used for itself, but also for irrigation and cleaning in the pavilions around. The roof of China Pavilion is equipped with a rainwater collecting system, from which the collected natural rainwater will also be applied in irrigation and road cleaning. By introducing the technology of small-scaled constructed wetland in the south landscape design of Region Pavilion, with the self-cleaning capability of the artificial wetland, an ecological wetland is created in the city center without covering a large number of land area.

Shanghai Corporate Pavilion chooses the environmentally friendly building materials (Fig. 2.127), adopts the world's latest solar power technology, and also equips itself with the rainwater collecting system. The collected rainwater will be

Fig. 2.127 Rainwater collection system in Shanghai Corporate Joint Pavilion

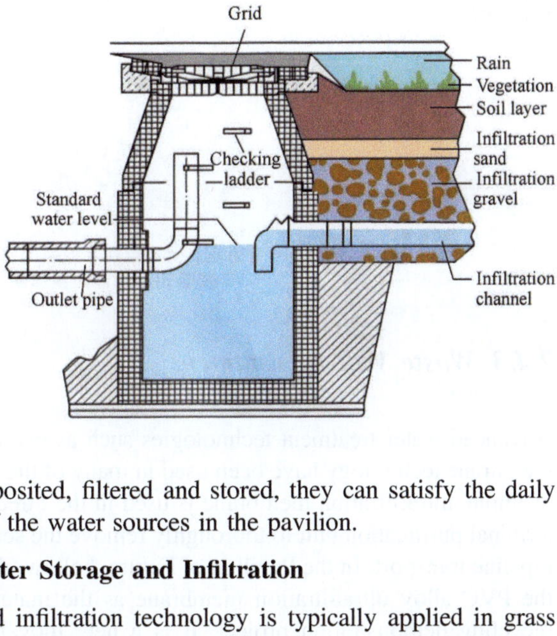

Fig. 2.128 Rainwater storage and infiltration solution for low green land and underground drainage system

recycled, and after being deposited, filtered and stored, they can satisfy the daily water consumption as one of the water sources in the pavilion.

(2) **Application of Rainwater Storage and Infiltration**

The rainwater storage and infiltration technology is typically applied in grass infiltration swales, lower green land, permeable road surface, potential infiltration wells and infiltration trench, etc. Among them, green land is a natural infiltration facility, which has good water permeability, saves investment and facilitates the absorption of rainwater. Figure 2.128 shows rainwater storage and infiltration solution for low green land and underground Drainage system. Figure 2.129 shows permeable pavement in the Expo site. Figure 2.130 shows typical structure of permeable pavement in the Expo site.

Fig. 2.129 Permeable pavement in the Expo site

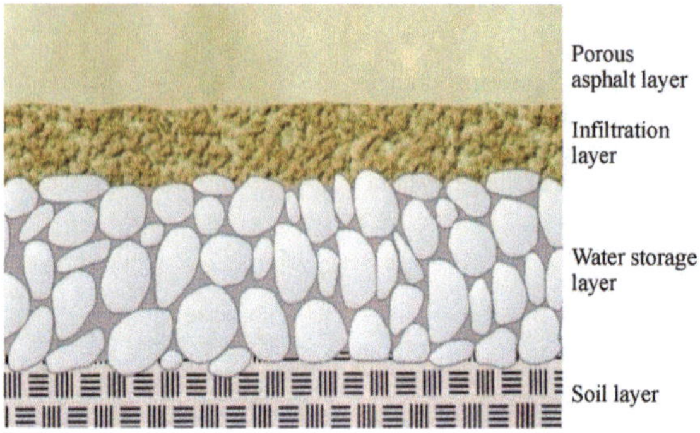

Fig. 2.130 Typical structure of permeable pavement in the Expo site

2.4.3 Waste Water Treatment

Advanced water treatment technologies such as reverse osmosis and ultrafiltration membrane technology have been used in many of the Shanghai Expo pavilions. For example, ultrafiltration membrane is used in the direct drinking water device as its terminal purification unit to thoroughly remove the secondary pollution in the water pipeline transport. In the Pavilion of Future, hollow fiber membrane bioreactor with the PVC alloy ultrafiltration membrane as the material is applied in the sewage recycling demonstration project. It is a new movable sewage treatment device, which collects the sewage from discharge and rainwater from the roof for reuse.

The research worked out the number of cases of the water purification, water collection and other applications of the water treatment in the Expo site, as shown in Fig. 2.131.

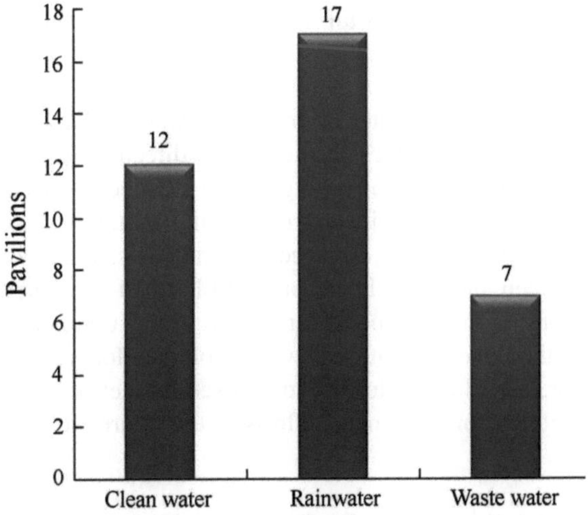

Fig. 2.131 Water treatment technologies applied in the Expo site

2.4.4 Summary of Water Treatment Technologies Applied in the Shanghai Expo

2.4.4.1 Current Status of Water Treatment Technologies in the Shanghai Expo

Shanghai Expo has given a full play to its concept of water saving, separate collection, assorting treatment and recycle in the process of water treatment and reclamation. It has adopted the foreign advanced technology of comprehensive rainwater harvesting system, and displayed successful cases of biochemical and advanced sewage treatment, in which the technology of membrane filtration, ultrafiltration membrane bioreactor, activated carbon adsorption, ultraviolet disinfection are applied. Nowadays, people have grasped the technical principles and practices of these advanced technologies, but problems such as high production cost, difficulty in regeneration, secondary pollution and short life circle have seriously hindered the promotion and application of these advanced technologies.

Due to the space limit of the Expo site, there are very little exhibitions for some special treatment of industrial sewage (especially the complex sewage that is difficult to be degraded), such as physicochemical pretreatment, biochemical intensified treatment, advanced treatment and water reclamation technologies. Along with the development of industry and people's greater awareness and requirements for environmental protection, a number of new technologies, innovative equipment and process have emerged for treating the industrial waste water that is difficult to be degraded.

2.4.4.2 Developmental Trend of Water Treatment Technology in the Expo

The concept of water saving and sewage reduction, separate collection, assorting treatment and recycle is the overall direction for carrying out the development of water treatment. For a long period of time in the near future, we should mainly focus on the combination of existing routine technologies and at the same time, perfect the existing advanced technologies. In response to the acute contradiction between increasingly serious problems of environmental pollution and people's improving requirements for protecting environment, we should place an emphasis on the development of new technologies for treating industrial sewage, in particular, the difficult-to-degrade mixed wastewater. And this will be the guiding direction for us to make efforts in the future water treatment.

2.4.4.3 Suggestions for Promoting Water Treatment Technologies in the Expo

Firstly, some relevant policies and rules should be formulated at a state level, so as to lead and help the development of new technologies in the industry and render great service to the society.

Secondly, the establishment of technical norms and standards should be set up in treating industrial wastewater, especially the difficult-to-degrade mixed wastewater.

2.5 Solid Wastes Disposal

The problem of solid waste is a growing concern in the development of the city and also the hot spot in resource recycling. The Shanghai Expo has focused on the demonstration of new technologies in the treatment and disposal of the engineering waste and urban solid waste, transforming the organic and inorganic garbage in the building and operation of the Expo Park into the construction materials, organic fertilizer and methane, etc. to achieve the harmlessness and recycling of garbage disposal.

2.5.1 Introduction to Solid Waste Disposal

Solid waste is the solid and semi-solid substances that we consume and discard in our daily life, production and other activities, more commonly known as trash or garbage. It consists of solid particles, trash, furnace slag, sludge, discarded goods, worn appliances, spoil, food scraps, human feces, etc. Some countries also classify the waste acid, alkali, oil and waste organic solvent into the solid waste.

2.5.1.1 Classifications of Solid Wastes

According to the Law of the People's Republic of China on Prevention of Environmental Pollution Caused by solid Waste, solid wastes consist of urban solid wastes, industrial solid wastes and harmful wastes. Based on the sources of the solid wastes, it is divided into the municipal solid wastes (MSW), industrial solid wastes and agricultural solid wastes.

(1) **Municipal solid waste (MSW)** commonly known as trash or garbage, is a waste type consisting of everyday items we consume and discard. It predominantly includes household waste, hospital trash, commercial wastes, building wastes (trash). Generally speaking, the daily trash generated amount per person is 1–2 kg, depending on the living standard, habit how they the discarded goods are recycled and municipal construction, etc. It is estimated that in 2025, 60 % of the global residents will live in the city or its neighboring area, the pollution caused by the MSW will become one of the serious problems. Normally, the higher the living standard, the more trash amount. In a industrialized city, the daily generated trash is 1 kg per person.
(2) **Industrial solid waste** is a type of waste produced by industrial activity, such as that of factories, mills and mines. Much industrial waste is neither hazardous nor toxic, such as waste fiber produced by agriculture and logging.

Based on its sources and physical properties, industrial waste is divided into six types. Based on the toxicity of the waste residue, it is divided into toxic waste and non-toxic solid waste.

Toxic waste, chemical waste, Industrial solid waste and MSW are designations of industrial waste. Sewage treatment can be used to clean water tainted with industrial waste. The wastes containing fluorine, mercury, arsenic, chromium, lead, and cyanogens elements or their chemicals and phenol, radioactive substances are toxic waste.

(3) **Agricultural waste** is also called agricultural garbage, mainly coming from feces and plants straw.

Solid waste disposal usually refers to physical, chemical, biological, physico-chemical and biochemical methods used to convert the solid waste into another form which is suitable for transportation, storage and utilization. The goal of solid waste disposal is harmlessness, minimization and reclamation. Some people think that solid waste is the most difficult to dispose among the "three wastes," because it contains rather complicated elements, with its physical properties (size, liquidity, uniformity, crushing degree, moisture, heating value, etc.) variable, making it hard to achieve the goal. Currently the main methods for proposal include compaction, crushing, sorting, curing, incineration and biological treatment.

Today the disposal of wastes by land filling or land spreading is the ultimate fate of all solid wastes, whether they are residential wastes collected and transported directly to a landfill site, residual materials from materials recovery facilities (MRFs), residue from the combustion of solid waste, compost or other substances from various solid waste processing facilities. A modern sanitary landfill is not a dump; it is an engineered facility used for disposing of solid wastes on land without creating nuisances or hazards to public health or safety, such as the breeding of insects and the contamination of ground water.

A landfill may offer a cheap solution to large amounts of trash, but incineration can quickly reduce the volume. An efficient and environmentally sound incinerator does not simply burn whatever trash there is lying around. Some incinerators recycle and reuse refuse as fuel to burn non-recyclable materials.

Modern incinerators are designed to destroy at least 99.9 % of the organic waste material they handle. Numerous thermal processes recover energy from solid waste. Companies burn in-plant wastes in conventional incinerators to produce steam. Pyrolysis, a process of chemical decomposition, produces a variety of gases and inert ash. Garbage burned in incinerators has poisoned air, soil, and water, which is one of the disadvantages.

2.5.2 Applications of Solid Waste Disposal Technologies in the Shanghai Expo

In the Expo site, set up an advanced infrastructure and services on solid waste collection by classification, transportation, processing and utilization of renewable energy, to achieve the recycling and utilization of the solid waste treatment.

In the core area of "One Axis and Four Pavilions" in Putong, the advanced foreign technology for pneumatic conveying was introduced and applied in large exhibition hall and municipal roads (Fig. 2.132). It mainly provided service to the "One Axis and Four Pavilions" (World Expo Axis, China Pavilion, Theme Pavilion, Expo Center, Performance Center) and the surrounding main roads.

Fig. 2.132 Pneumatic conveying trash bin in the Expo site

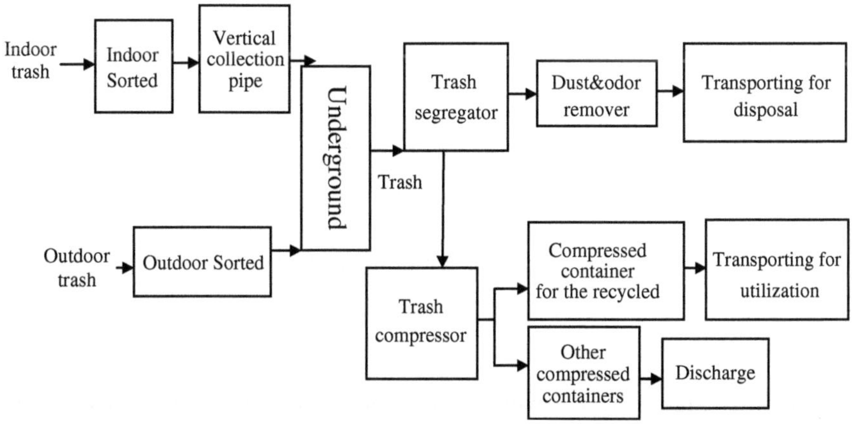

Fig. 2.133 Flowchart for pneumatic conveying trash system

This system is a closed pneumatic system for trash transmission and the whole collecting procedure is pre-programmed by the computer. The transportation of trash is completed in a totally closed status without any manual operation. The sanitation workers can avoid the direct contact with trash so as to ensure that the trash disposal process emit no smell or overflows. Figure 2.133 shows flowchart for pneumatic conveying trash system.

2.5.2.1 Case Demonstration

In Basel, Geneva and Zurich United Case Pavilion, one can see the very fine sorting of trash. Even the glass bottles are subdivided into several groups like brown, green and colorless ones, which contributes to the recycle and disposal of trash. In Sweden, the residents have the obligation to recycle waste, for example, diving enthusiasts will go to the Rhine River in search of wastes every year, the students are organized every year to do some environmental protection work within their power, such as picking up litter or garbage sorting.

While the Porto Alegre Pavilion demonstrates a creative way to transform trash into things of value—to extract fiber from PET bottle to make bedspreads and pillows—this brings income and job opportunity for women from 20 poor families. And this project also improves the living quality for more than 90,000 people in need.

The exhibiting project in Cairo Pavilion involves the transformation of an 80-acre dump of the 1980s into a park.

Taipei Case Pavilion shows their approach in proper classification and recycling of trash, to not only reduce the amount of trash but also realize “zero landfill.” In 2000, the daily amount of trash produced by all households in Taipei was 2,970 tons; while in 2009, the number was reduced to 1,009. In other words, the

Fig. 2.134 Work schematic for Sysav garbage power plant

amount was reduced by 2/3 within only 9 years, the success of which can be attributed to the policy of "no trash on the floor." There's much in their practice that we can learn and draw experience from, and it is advisable to promote this kind of practice in mainland China.

Speaking of the four-floor small building of Shanghai Pavilion, we can see that the constructional materials of this building are actually made of town refuse. During the reconstruction of some old buildings in the city of Shanghai, about 150,000 pieces of bricks and some structural steels were left over, which are now re-used in the construction of the walls and stair treads of this building, and are re-welded into its steel stairs.

Malmo Case Pavilion of Sweden demonstrated how Sysav Trash Disposal Company makes flexible utilization of the wastes to convert them into source of energy. Every year the company has to dispose the MSW and industrial waste in a city of 620,000 residents. It collects the flammable garbage of up to 400,000 tons a year in the southern area of Sweden, after combustion to produce 1,000 GWh (1 billion kWh) of electricity, equivalent to 300,000 tons of standard coal production. The latest production line generates 135 GWh of electricity and 540 GWh of hot water for district heating, which meet the heating demand for 40 % of Malmö and Burlöv area (300,000 residents) and electricity demand for 170,000 households. The new production line is capable of handling 200,000 tons of combustible wastes per year, running 24 h a day, 365 days a year. Figure 2.134 shows work schematic for Sysav garbage power plant.

2.5.2.2 Conveying Trash Out for Disposal and Utilization

(1) **Landfill of Household Trash**

The seepage-proof system with high safety is applied to prevent the waste from leaking. The seepage-proof system of landfill site has used foundation, woven geotextile, metalling, nonwoven geotextile, rough-surfaced geomembrane of 1.5 mm thick, and many other materials to ensure its excellent performance, so that trash and underground water is completely insulated without even the smallest gap in between.

(2) **Incineration of Household Trash**

The factory for incineration of household trash spent 70 % of their total investment on the gas control system. The world's most advanced technology of "lime + activated carbon + bag filter" is applied in the gas purification process.

(3) **Utilization of Building Trash**

Over 60 % of the road surface in the Expo is made from building trash and slag. And most of the pavilions in the Expo have adopted recyclable building materials.

The construction wastes in the Expo site is first classified according to their uses, of which 10 % is of higher economic value to be sold off, the remaining 90 % is waste concrete, waste plastics and waste timbers. What is the destination of these construction wastes? The answer is: construction waste residue with high water resistant strength is used as the road base in the Expo site; construction wastes recycled aggregate as roadbed construction; waste concrete is processed into pavior bricks; waste wood and plastics is used as inspection well lid, inside and outside the Expo site. The smooth implementation of the project has given people an inspiration. In the urban construction, is there any possibility of establishing such a large-scale "Logistics Transformation Center" for the construction wastes?

Chapter 3
Typical Low-Carbon Pavilions in the Expo

During Expo 2010, pavilions of different countries have not only demonstrated the theme of "Better city, Better life" to the public, but also given a full play to the great efforts they have made in achieving the reduction of carbon emission, energy saving, and maintaining a sustainable development.

3.1 One Axis and Four Pavilions

The Expo Axis, China pavilion, Theme Pavilion, Expo Centre and Culture Centre are collectively called "One Axis and Four Pavilions" with a total construction area of 780,000 m^2 (Fig. 3.1).

The "green achievements" which are applied in the four pavilions along the axis include many energy-saving technologies, which are both environment-friendly and ecological, such as solar energy, LED lighting, ice storage, ground source heat pump, roof rainwater harvesting and utilization, river water circulation cooling, pneumatic recycling, water-saving irrigation for greenbelt, and so on.

The main pavilions including Theme Pavilion, Expo Centre, Nanshi Power Plant as well as China Pavilion have generously offered their roofs and walls to establish a largest building-integrated photovoltaic grid power system in China or even in Asia. "Solar power," with a total installed capacity of 4.5 MW, is expected to reduce carbon dioxide emissions by 4,235 tons. Among those venues, the solar panels installed in Theme Pavilion are up to 30,000 m^2 and are known as the world's largest solar roof in single area with an annual generation capacity of over 2.8 million kWh, which is to reduce 2,500 tons of carbon dioxide emissions per year. Figure 3.2 shows PV integration system in the Theme Pavilion.

The roof of the Theme Pavilion has double layers, with the lower layer is designed to solve the functional problems such as construction water proofing, insulation and lighting; the upper layer is in diamond shape, installed with both the light-proof solar PV module and translucent solar PV modules, making a perfect combination of architecture and functions.

UNIDO-ISEC and Jiangsu Modern Low-carbon Technolgy Research Institute (eds.),
The Research Report on Application of Low-carbon Technology in Expo 2010 Shanghai,
DOI 10.1007/978-3-662-44357-6_3

Fig. 3.1 Panorama of "One Axis and Four Pavilions"

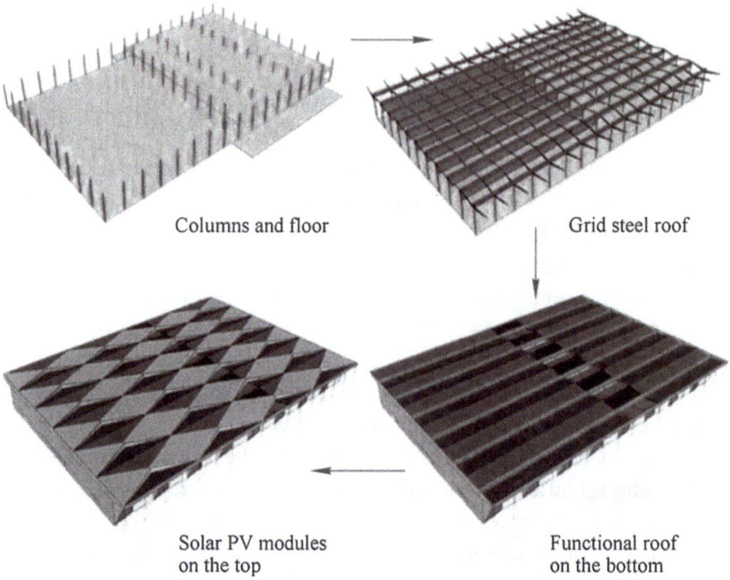

Fig. 3.2 PV integration system in the Theme Pavilion

The skylight of Theme Pavilion can both introduce the sunshine in and make the room airy. The roof rainwater collection system is used for plants irrigation, and meanwhile takes away the heat of the building, so that during the daytime no high-powered air-conditioner is required. The pavilion also has a largest ecological wall of 6,000 m^2.

The roof area of the Expo Centre is 40,000 m^2, with a total installed capacity of the solar system on the roof and wall reaching 1 MW. Figure 3.3 shows solar energy utilization in Expo Centre. Figure 3.4 shows solar sun-shading system on the roof of the Expo Centre.

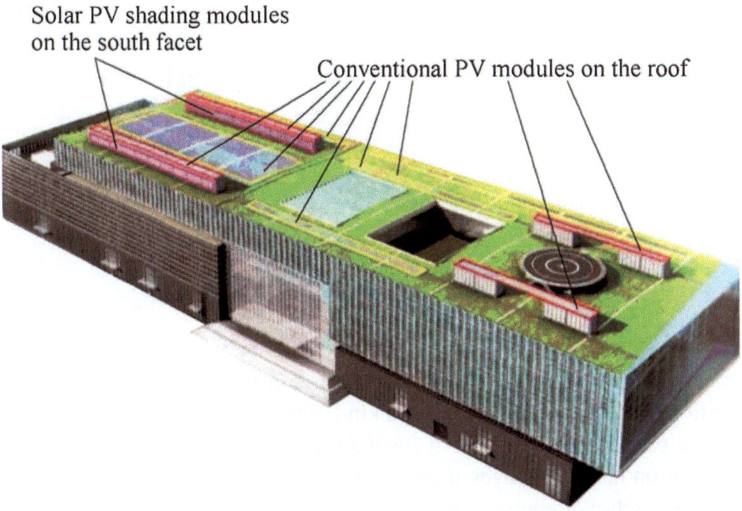

Fig. 3.3 Solar energy utilization in Expo Centre

Fig. 3.4 Solar sun-shading system on the roof of the Expo Centre

The curtain wall system of the Expo Centre has good thermal performance with dual silver silk low-E insulating glass in the insulating layer by filling argon to improve the thermal insulation properties. At the same time, the Expo Centre also features a series of advanced technologies such as ice storage air-conditioning, river water source heat pump, heat storage solar hot water system, rainwater control and utilization system and water collection and use systems, variable frequency water supply system, computerized irrigation system for greenbelt, the central monitoring and management systems and energy-saving equipment, etc.

Among these, the material used in constructing the exposed wall of China Pavilion is nonradioactive and pollution-free green product. For instance, all the

Fig. 3.5 Bird's eye view of China Pavilion

doors and windows adopt low-E glass, which can reflect heat and reduce energy consumption. Besides, it can also spray a kind of painting, which can change sunlight into electric power and then store it for lighting the exposed wall. The platform of Region Pavilion has been covered with a soil layer of 1.5 m, which can save above 10 % of the pavilion's energy consumption. The observatory on the top of Nation Pavilion may bring into the most advanced solar energy film to store and change sunlight into electricity. On the top floor, there is also rainwater collecting system, by which the rainwater is purified and used to wash the washroom and the automobiles. The bracket floor of the main construction forms a self-sun-shading structure, which has already provided shelter for the lower space from sunshine and therefore saved energy. Ice storage technology achieves peak load shifting of electricity. River water cooling loop increases the efficiency of the air-conditioning system and significantly saves the electricity. Figure 3.5 shows bird's eye view of China Pavilion. Figure 3.6 shows solar technology utilization on the roof of China Pavilion.

In the Expo Axis (Fig. 3.7), the enormous air-conditioning system is not driven by electricity, but by the natural water source and ground source of the Huangpu River, of which water source accounting for 70 % and ground source 30 %. In summer, water source and ground source heat pumps function as hot and cold

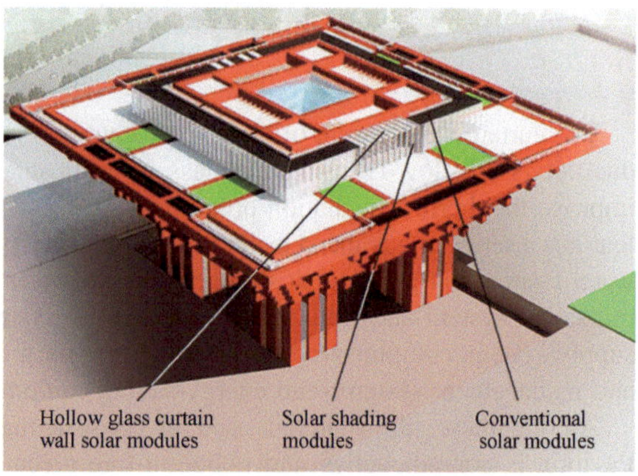

Fig. 3.6 Solar technology utilization on the roof of China Pavilion

Fig. 3.7 Panorama of the Expo Axis

sources for air-conditioning system so that no need for cooling tower to supply water. In winter, both the river and geographic radiators can be used as hyperthermia heat source and cryogenic heat source of the heat pump system. The Sunshine Valley of the Expo Axis implies an important scientific idea, namely creating a comfortable green underground space. It not only "collects" sunshine from the 40-m-high altitude to the underground, but also conveys the fresh air to the underground, alleviating the depressing effect of the underground and achieving the goal of energy conservation. In addition, rainwater can flow to the gutter waterway into the second floor of underground along the glass curtain walls and then joined to the impounding reservoirs of 7,000 m^3 to be recycled after a series of disposal. The recycled water can be used for irrigation to the sinking types of gardens on both sides and washroom flushing. Through the Sunshine Valley, the total amount of water consumption in the Expo Axis can cut down by 50,000 m^3.

In terms of building energy efficiency of the Expo Cultural Centre (Fig. 3.8), their approach is by efficiently controlling the ratio of walls and windows to an extent, and designing overhanging structure to provide sun shading. Meanwhile, many technologies such as river source heat pump, ice storage system, pneumatic trash collecting system, air-conditioning condensed water and roof rainwater collecting system, and program-controlled water-saving irrigation system, etc., have been used.

LED is used in all the landscape lighting in the One Axis and Four Pavilions, which is first used in the blocks so extensively.

Fig. 3.8 Expo Cultural Centre

3.2 Japan Pavilion

The solar batteries installed in the exterior walls of Japan Pavilion (Fig. 3.9) enable the walls to produce energy on their own, which for the first time in the world adopts the technology of electricity-generating film. Moreover, Japan Pavilion has applied another technology called "circular breathing pillar," which can take in sunshine, store rainwater, and absorb fresh air, and it is also a pioneering project in the world.

Within Japan Pavilion, the vertical circular breathing pillars are installed under the floor and in other positions. The pillar is actually an environmental circular system integrated with the inner structure of the pavilion's construction. It can help relieve the burden of environment by collecting sunshine, gathering rainwater for roof spray, and keeping air ventilation. Electricity-generating film is a highly transparent thin film with solar batteries inside which has good transmittance and generates electricity. In the waiting area inside the pavilion, through artificial

Fig. 3.9 Panorama of Japan Pavilion

spraying, the surrounding heat is absorbed in the process of vaporization, and thus makes a cool environment for the tourists.

The construction of the pavilion combines the cutting-edge environmental control technology with material technology, makes full use of nature, so as to alleviate the burden of environment. In constructing Japan Pavilion, the application of electricity-generating film and circular breathing pillar helps substantially reduce the steel consumption to 60 % of a normal building.

The "Breathing hole" and "Antenna" of the purple silkworm island not only endows the pavilion with life but also bears many functions. It is the core structure of cooling system and airing system in Japan Pavilion. The "Breathing hole" and "Antenna" are actually six circular breathing pore canals erected in the pavilion, connecting the roof above and the basement on the bottom (Figs. 3.10 and 3.11).

Fig. 3.10 Antenna of Japan Pavilion

Fig. 3.11 Breathing holes on top of Japan Pavilion

On rainy days, the sunk "Breathing hole" introduces the rainwater to the underground tank through those breathing pillars. After the water treatment, the reclaimed water is obtained, supplying to the water-spraying system outside the pavilion to cooling the pavilion. The "Antenna" is the vent tower of the pavilion, which built on the "Breathing hole." The breathing pillars, built from transparent materials, function as the light columns in daytime when sunlight are shine in, and thus provide light source for the lower space in Japan Pavilion.

Without any piles, the construction of Japan Pavilion combines concrete and natural soil to form a firm foundation, not only improving the bearing capacity of the land but also protecting the earth's surface. Grass and trees can be planted there after digging up the foundation soil when the pavilion is disassembled.

3.3 Switzerland Pavilion

Covering an area of 4,000 m^2, Switzerland Pavilion (Fig. 3.12) demonstrates the glorious beautiful prospects for the future. With sustainable development as the core value, its design fully shows how to combine nature with high-tech elements. Its external curtain, made from degradable soybean fiber, is able to generate electricity. Overlooked from the sky, its outline is like an imaginary vision for the future world.

The design concept is visible in a mix of many contradictory elements, the Chinese Philosophy of "Yin" and "Yang" is well reflected in the entire building, to achieve an ingenious combination of high-tech and nature. Cable car is also put into its design so that visitors can get to the roof for sightseeing. The chair lift will lift visitors from the crowded city to the roof where a tranquil space close to nature and far away from the crowd is provided.

Fig. 3.12 Panorama of Switzerland Pavilion

3.4 Italy Pavilion

The design of the Italy Pavilion (Fig. 3.13) is inspired by a traditional children's game "pick-up sticks," which comes from Shanghai. The rectangular pavilion has been laced with intersecting lines—representing pick-up sticks. It comprises 20 free assembling functional modules of different shapes, bounded by the "sticks." They represent Italy's 20 regions.

The whole pavilion embodies a mini Italian city. Strolling in the pavilion, visitors may feel like walking in a city boasting narrow streets, plazas, courtyards, and lanes with an Italian flavor. The country's achievements in technology and environmental protection will also be displayed.

The pavilion adopts solar energy technology and air flow and water flow in summer to lower the temperature, which is novel in design. Besides, it uses the latest invented diversified material—transparent concrete. This type of concrete, added with glass, has different degrees of transparency, able to perceive external temperature and humidity of the building. The blade-shaped internal design can not only produce imaginary shadow effect, but also import natural wind into the pavilion. The idea of environmental protection and sustainable development has been paid great attention to since the very beginning of the design.

Italy Pavilion puts forward the strategy of eco-climate, which aims to lower temperature by solar radiation in winter and air–water flow in summer. Hot wind can be eliminated through automatic regulation system, thus lowering the internal temperature of the building. Along with radiation control, heat can gather on the transparent glass containing photoelectric integrating modules so as to fully save electric energy. Lastly, the design of the top can effectively prevent the roof from rainwater erosion.

Fig. 3.13 Italy Pavilion

3.5 Finland Pavilion

Finland Pavilion (Fig. 3.14) is nicknamed "Kirnu," shaped like a curling floating on the lake. The most innovative construction material of the pavilion is a composite UPM ProFi—shingle based on recycled paper and plastic. 25,000 injection-molded marble white scaly shingles cover the outer facade of the pavilion. These oddments from tagboards and plastics are solid and wear-resisting on the surface, with low content of moisture and light weight, will not fade.

The bowl-open design on the top of the "Kirnu" can facilitate natural ventilation and easy for laying the solar cell panel to supply electric power for cooling devices inside the pavilion. It can also collect rainwater for recycling. In addition, the "Kirnu" is designed according to standard of permanent construction so that it can be easily demolished after the World Expo and reconstructed somewhere else in the future.

3.6 Canada Pavilion

The building of Canada Pavilion (Fig. 3.15) features the redwood-clad facade. The redwood used is specially grown in Quebec City, Canada, and transported to Shanghai by air. Not susceptible to erosion, the redwood will not become mildewed even in the damp city of Shanghai and has good performance in warm keeping and sound proofing.

The technologies in recycling are reflected in the construction of Canada Pavilion. Its exterior walls are covered by vegetation—a special greens. Rainwater is harvested and utilized through the drainage system. There are no large-sized exhibits or articles on display to ensure that the air inside is well ventilated. Meanwhile, the barrier-free design enables anyone venturing inside the pavilion will be able to ride bikes that ate flanked on all sides.

Fig. 3.14 Finland Pavilion

Fig. 3.15 Canada Pavilion

3.7 Hamburg Case Pavilion

The Hamburg House, which is known as the first Passivhaus in China, covers a net floorage of 2,300 m^2. Its design proposal is based on the prototype of Hafen City in Hamburg. Giving full consideration to the climate in Shanghai, it adopts the principle of energy saving in accord with the construction of Passivhaus. By using energy-efficient technology, the pavilion uses as little energy as possible while still ensures the comfortableness inside the room, cutting down the heating demand by 15 kWh/(m^2a) and keeps the primary energy consumption to less than 50 kWh/(m^2a).

The pavilion can keep room temperature at 25 °C all year round, and its energy use is only 10 % that of an ordinary building.

The Passivhaus achieves its good effect of heat insulation by frequent ventilation rate, so the pavilion does not have to use a traditional heating device. The reason we call it "passive" is that the main thermal energy of this kind of building comes from the sun, human body, and the remaining heat of electric equipment and it consumes very little energy resources. Besides, the construction of such architecture is not restricted by the types of different buildings; any type of building can be reconstructed to meet the standards of Passivhaus.

This architecture features solid structure with compact volume. The exterior protected construction has good performance in heat preservation and air impermeability, which consequently reduces the energy consumption of heating and cooling in winter and summer. The passive energy resources from the sun, human body, and indoor electric equipment can meet the majority demand of heat. Besides, the central ventilation device can provide all the rooms in this building with the dehumidified "new" wind, which has been heated or refrigerated. Figure 3.16 shows view inside Hamburg Case Pavilion. Figure 3.17 shows visitors' experience in Hamburg House.

The energy resources for heating and refrigerating of this building come from the hot and cold water produced by a ground source heat pump device. The foundation pile, which is 35 m deep underground, is connected with the underground pipe

Fig. 3.16 View inside Hamburg Case Pavilion

Fig. 3.17 Visitors' experience in Hamburg House

network to collect geothermy. The heating and refrigeration of the building mainly relies on hidden-pipe systems on each floor. The ventilation device that is used to recycle the heat can bring "new" wind to the rooms. The working rate of heat recycling of the central ventilation device that is installed on the roof can reach at least 90 %, and refrigeration rate is at least 80 %. Moreover, there is a PV power generation device that covers an area of 450 m^2 on the roof, and it can provide approximately 80 % of electricity for running and using the building.

3.8 London Zero-Carbon Pavilion

London Zero-carbon Pavilion (ZedPavilion) is the first architecture without carbon dioxide emission in Expo history. The building is designed by Zero-carbon Centre and supported by industrial and commercial circles and the government. Inside this pavilion, visitors can experience a real Zero-carbon life by touring the pavilion's Zero-carbon conference hall, restaurant, exhibition hall, and six Zero-carbon show houses of different styles.

The prototype of this pavilion is taken from "Beddington" zero energy development (BedZed) in London which was the first Zero-carbon emission community in the world established in 2002. Tens of thousands of people go to visit the community and learn from the Zero-carbon constructional technology along with the environment-friendly technologies and concepts of renewable energy resources. Figure 3.18 shows low-carbon technology integration in London Case Pavilion. Figure 3.19 shows technologies classification in London Case Pavilion.

One of the highlights of Zero-carbon Pavilion is the special fluorescence paint attached to the surface of the wall. The architecture uses the solar energy collected in the daytime and releases fluorescent light in the evening to reduce energy consumption of lighting, which makes the pavilion a shining house. Besides, the Zero-carbon pavilion also shows us "windows that can generate electricity" through a perfect combination of sunlight collection and electricity generation.

In the ZedPavilion, heat ventilation demand is provided by the river source public system and the solar wind-powered refrigerating air cap system. And the electricity is provided by solar panels and bioenergy units affiliated to the building, so as to meet its energy demand all year round, and provide visitors with a comfortable experience with the least energy consumption.

On the slope-shaped roof, there are different colors of wind cowls rotating flexibly along with the wind direction. It makes use of the pressure of temperature and wind to continuously bring in fresh air to different rooms so as to refresh and

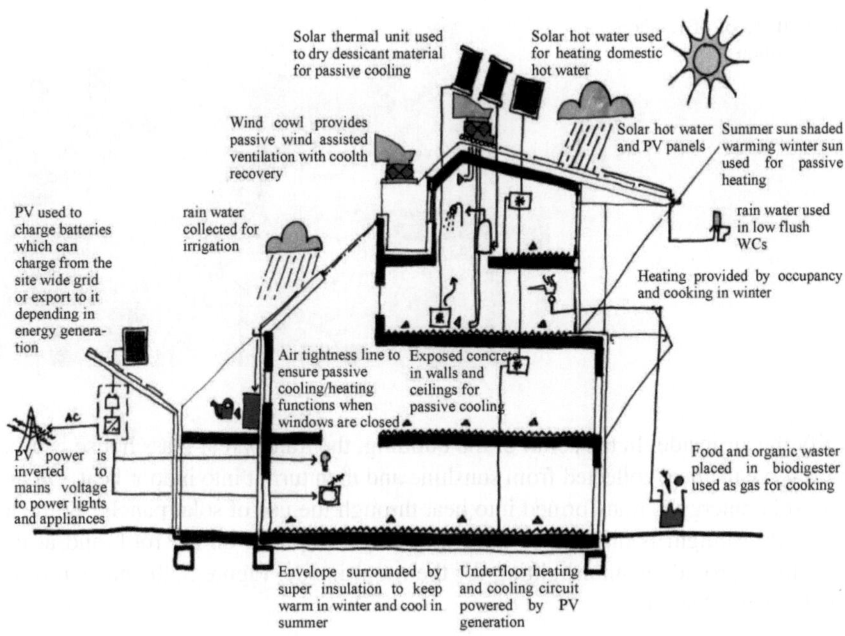

Fig. 3.18 Low-carbon technology integration in London Case Pavilion

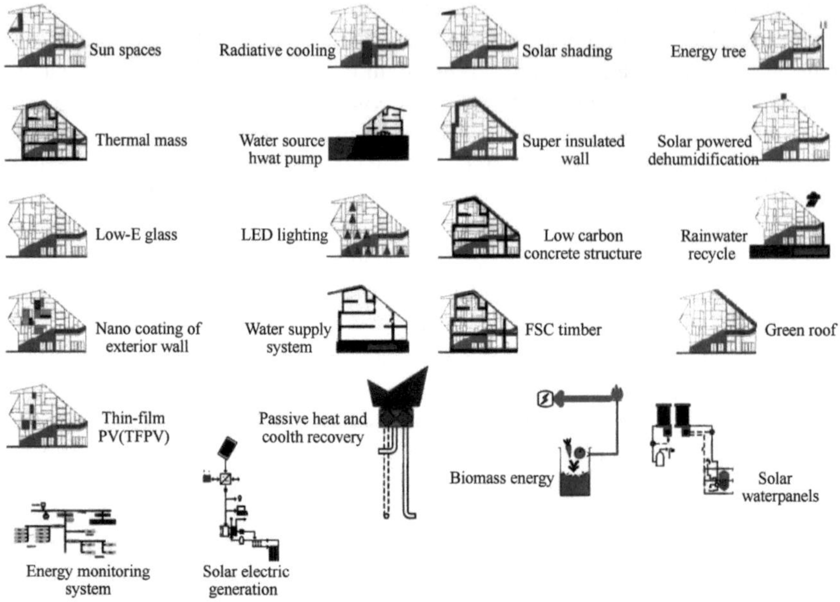

Fig. 3.19 Technologies classification in London Case Pavilion

Fig. 3.20 Roof of London Case Pavilion

purify the air inside. In the south of the building, the transparent glass house is used to preserve the heat collected from sunshine and then turn it into indoor heat. On the roof, solar energy is transformed into heat through the use of solar panels. While to the north, sunlight is diffused to cultivate the green plants on the roof, and at the same time, provide natural lighting for the room inside. Figure 3.20 shows roof of London Case Pavilion.

3.9 Ecological House in Shanghai Case Pavilion

The ecological house in Shanghai Case Pavilion is themed with "Ecological Construction, Blissful Life" and demonstrates connotative meaning of the green and low-carbon technology, as well as the principles of design based on local conditions (Figs. 3.21, 3.22, 3.23 and 3.24). It has a floorage of 3,017 m^2 with four floors above the ground and one underground. Integrating the four major technological systems, including energy saving, resource recycling, environment suitable for living, and residence intelligent and efficient, the construction of this ecological house has applied altogether ten major systems with unique characteristics. To be more specific, these systems are waste recycling concrete, exterior protected structure which can insulate heat and adapt to climate change, household fuel cell energy centre, BIPV electricity generation, natural ventilation technology, day lighting and LED indoor lighting, wall-hanging module afforesting, household trash

Fig. 3.21 Microalgae railing of Shanghai Eco-Home

Fig. 3.22 Vegetation on the roof of the Shanghai Eco-Home

Fig. 3.23 Side wall greening of the Shanghai Eco-Home

Fig. 3.24 Eco-Home in Shanghai Case Pavilion

recycling, intelligent integration management, household remote controlled medical care, and household robot service. The concept of recycling is displayed everywhere inside the house. For example, the water used to flush the toilet comes from rainwater. And there are two kinds of elevators in the house. One is used to collect

Fig. 3.25 Solar energy system on south sloped roof

potential energy during the ups and downs of the elevator. The other one can adjust its speed according to the number of people taking it.

The walls material of the pavilion comes from the 150,000 pieces of bricks left over during the reform of some old buildings in Shanghai. The majority of the bricks used inside the house is made from the sludge hollow bricks after watercourse dredging and wastes from the factories, so are the desulfurated plasterboards. Besides, the wooden surface of the house is actually made of pressed bamboos. Since bamboo has a short growth cycle and is easy to obtain, this saves the wood consumption.

In order to take advantages of draught in summer, the structure of the Eco-Home caters to the direction of dominant monsoon in this season, thus it faces south and is arranged in strips. The building is designed with many horizontal air ducts for the wind to run through and reinforce the ventilation. Five groups of vertical mute wind-generating sets are installed on the roof to provide green renewable resources, which can be directly used by the building. At the same time, luminous energy can as well be transformed into new resource, and the roof shutters which follow after sunshine can change their angles automatically according to the changing angle of the sun. On the one hand, they can be used for sun shading, and on the other hand, they can reflect the sunlight to improve the illuminance of indoor rooms. When indoor sunlight cannot perform its function of lighting, the curtain shutters will adjust themselves automatically, and the indoor lights will turn on at the same time.

The PV power generation system used on the south-faced sloped roof and balcony can produce clean electric energy. The solar water-heating system together with the PV power generation system can produce hot water by 900 L per day, which accounts for about 50 % of daily hot water consumption. Besides, there are also solar energy panels on the exterior walls. Inside the building, light guide system is used so that the rooms facing north can also capture the sunlight. Figure 3.25 shows solar energy system on south sloped roof.

3.10 Alsace Case Pavilion

Alsace Case Pavilion is surrounded by green plants. However, what is most surprising about this pavilion is not the green wall but its "waterfall solar wall" (Fig. 3.26). The "wall" consists of three layers. The exterior layer is solar panel and the first layer of glass, the middle is sealed cabin, and the third layer is waterfall glass.

Alsace Pavilion covers an area of 2,100 m^2 with four floors in total. It adopts solar panel technology on its exposed wall and most part of the pavilion is made of glass, with the solar panels stretching outside. The north and south exposed walls are covered with low plants, which have a natural function of heat insulation, and render a special outlook with the "waterfall." The building is designed by the eight design firms of Alsace Architecture Research Institution and has taken 4 years to finalize the design.

The solar panel architecture is one of the ten most creative architectures in France. Before transferring the whole design to Shanghai, the design team firstly considered the climate in Shanghai and used solar panels to develop two systems, one for heating and the other for refrigerating, respectively, applied in winter and

Fig. 3.26 Solar waterfall wall of Alsace Case Pavilion

summer. In summer, in order to deal with the hot weather, the designers use some jacks to support the solar panel and introduce into the water from outside to lower the temperature inside. While in winter, the “air” in between the glass is heated through sunshine and then imported to the indoor rooms. As a result, this system becomes an adjustable device which provides a cool summer and warm winter. Although Alsace is an industrial area, the local government has attached great importance to ecological and environmental protection, and has put the green constructional standards into practice since many years ago.

When the waterfall wall operates in winter, the seal cabin closes and the sunlight shining upon the photoelectricity panel will be transformed into electricity. Through the radiation of sunshine and the heat produced by the panels, the air left in the seal cabin between the two layers of glasses will be preheated and continuously provide heat for indoor rooms. This waterfall wall shows that a building itself can have a mechanism for climate adjustment, and it is especially helpful in those areas which are cold in winter and hot in summer.

Apart from the overall design, most of the exhibits inside the pavilion are associated with ecological environment, such as the most characteristic vineyards in Alsace, the innovative biochemical valley, the new ecological community, and light railways in Strasburg… What attracts most attention is the “frog pedaling water-wheel” on the ground floor. This cartoon image is designed by Tony Ungerer, a designer and writer, showing the device for generating electricity through water-power. In this specific case, the turning of the waterwheel, the spraying, and the mini energy-saving bulbs all completely rely on this waterpower device.

3.11 Broad Pavilion

As a sustainable construction, Broad Pavilion (Fig. 3.27) has taken to use the polymerization foaming material of 170 mm thick on its walls and roof to achieve heat preservation. As for the windows, three-layered glasses with plastic frame are

Fig. 3.27 Panorama of Broad Pavilion

Fig. 3.28 Energy-saving windows inside of the Broad Pavilion

used, with a function of automatic sun-shading outside and heat insulation inside (Fig. 3.28). And the ventilation system adopts "fresh-wind fan" which can reclaim heat to cool/heat the new wind by the removed cool/hot air. The measures taken to insulate heat from the aspects of walls, windows, and ventilation make this building not too cold even without heating equipment in winter. And the air conditioning in summer is only for human body, lighting and heat from electrical appliances, the energy consumption of which is only about 20 % of the amount of traditional buildings.

The pavilion is also a super clean building "capable of breathing." The well-ventilated indoor rooms can replace air thoroughly three times per hour, which is double of a five-star hotel. In particular, due to the use of the fresh-wind fan which has electrostatic dust collector, 99.9 % bacteria and dust inside the air will be filtered out when the outdoor air enters in through the device.

In regard to energy efficiency in the construction, compared to the same type of the building, Broad Pavilion has saved the raw materials of six times, energy of five times and reduced the construction wastes by 300,000 tons.

3.12 Madrid Pavilion

Step into Madrid Pavilion (Fig. 3.29), visitors can see a rectangle four-floor building of 18 m high, which looks like a giant box wrapped tightly by bamboos. The pavilion has kept the external features of a local bamboo house in Madrid, and uses bamboo slices to cover the house surface so as to create an original ecological landscape. Bamboo slices have gaps in between, which can adjust beams, insulate heat and noise to make the room warm in winter and cool in summer. In order to better adapt to the climate and environment of Shanghai, one side of the bamboo house is not wrapped with bamboo slices, instead, it adopts the energy-saving glass curtain wall to make full use of natural lighting. The solar energy was intended for the rooftop of the bamboo house, but considering that the sunlight intensity in

Fig. 3.29 Panorama of Madrid Pavilion

Fig. 3.30 Solar energy devices and greening on the roof of the Madrid Pavilion

Shanghai is not as much as that of Spain along the west coast of Mediterranean, the designers use solar energy and electricity as the energy source of the bamboo house (Figs. 3.30 and 3.31).

"Air Tree," a unique landscape in the avenues of Madrid, is built in Madrid Pavilion. It is an imitation of one of the "Technological Trees" along the Eco-boulevard in Madrid, which aims to demonstrate the technology of wind power. Inside the air tree, there are a series of small impeller generators and large fans to lower the temperature inside the construction (Fig. 3.32). It works like a programmed robot, moves according to the Sun's position change, and transforms its external shape along with the sunlight intensity. Besides, there is an advanced audio-visual screening system inside the air tree. The screen of the system is actually a three-layered curtain that has wrapped up the internal structure of the air tree. The external surface of the air tree is covered by a light-permeable membrane, whose internal area reaches about 300 m^2. As an open public space, the air tree provides the audience with a place to walk and relax freely.

Fig. 3.31 Close shot of "Bamboo House" in Madrid Pavilion

Fig. 3.32 Internal system of air trees

Air tree adopts the environmental technology of sun-shading and heat-insulating to provide visitors with a comfortable environment. On the rooftop of the air tree, the solar batteries and shutters which can open automatically are installed to provide itself with needed energy. Considering the temperature during the Expo, designers fill up inside the air tree with plants, and at the same time, install a draught fan with a diameter of 7 m to ventilate the indoor rooms. Therefore, without air-conditioning system in the air tree, visitors can still feel cool in the room where the temperature inside is at least 3–5 °C lower than that outside in hot summer.

3.13 Shanghai Corporate Joint Pavilion

On the roof of Shanghai Corporate Joint Pavilion, some 2,200 m^2 of solar thermal collector is installed to collect the 95 °C hot water generated by solar energy, and through the new technology of ultralow-temperature power generation, the output electric power can exceed over 200 kW. This technology opens up a new way to generate electricity through solar energy. And the electricity generated can be used for the exhibition of architectures and in daily household. When this “Magic Cube” is lit up by LED lights during the day and night, bright spots and images are flickering around the building, with the supporting energy coming from the infinite space.

As for the façade of the pavilion, the polycarbonate transparent plastic tube is adopted, integrating different kinds of technical equipment pipes, to create an illusory and vague image on the exposed vertical wall. After the World Expo, these plastic pipes can easily be recycled so as to reduce energy consumption for the whole society.

The rainwater within the pavilion can also be reclaimed, and through the processing technology of precipitation, filtration, and storage, it can help meet the daily water demand inside the pavilion. Besides, it can also be used as a “spray,” which not only lowers the temperature nearby, but also purifies the air to bring a comfortable microclimate within a small space.

Chapter 4
The Concept of Low-Carbon Life in the Expo

To embed the low-carbon concept into people's life is much more important than to popularize the technology. There is no doubt that the research and development of low-carbon technology as well as its popularization is of great significance, but if the green ideas of protecting and caring for environment is not deeply rooted in people's mind, such prospects as low-carbon economy and low-carbon life are nothing but an illusion. That is why most of the things on display in developed countries' pavilions are more about how to help people actually internalize the low-carbon concept.

In this Expo, the green ideas have actually touched every aspect of life, including the articles of everyday use such as staff working clothes, furnishings, tableware, and the like, which interprets the concept of "Saving Creates Better Life."

4.1 Low-Carbon Garments

The green or blue color clothes that Expo volunteers wore as uniforms have many functions, such as rain-proofing, wind-break, dirt-resistant, and moisture-absorbing (Fig. 4.1). Besides, they can fend off ultraviolet rays, retard flame, and dry quickly. In addition to these functions, this kind of material can also cool down instantly. When touching it the skin can instantly have a cool feeling that the temperature is lowered by 1–2 °C. While exposed to sunlight for a long time, the skin can sense the temperature drop of up to 3–5 °C.

The uniform that the staff in Germany Pavilion wore was made of a kind of ecologically cycled fiber (Fig. 4.2). This special Dacron fiber can be recycled afterwards.

In Japan Industry Pavilion, the Kimono worn by the staff was made of a new kind of recyclable material, which is woven by heat-resisting bio-plastic fiber and a kind of Kyoto traditional silk fabric (Fig. 4.3). After the Expo, the uniforms, together with other equipment were sent back to Japan, put through a recycling system, and used for reproduction.

UNIDO-ISEC and Jiangsu Modern Low-carbon Technolgy Research Institute (eds.),
The Research Report on Application of Low-carbon Technology in Expo 2010 Shanghai,
DOI 10.1007/978-3-662-44357-6_4

Fig. 4.1 Expo volunteers

Fig. 4.2 Staff uniforms in Germany Pavilion (*Source* World Expo Bureau)

4.2 Milk Carton Benches

In the Expo site, there were 1,000 benches made of recycled paper milk cartons (Fig. 4.4). Each bench was 1.2 m long and 0.4 m wide. Although they look ordinary in shape, these benches are resistant to sunshine and rainwater. The commonly seen

Fig. 4.3 Working uniforms in Japan Industry Pavilion

Fig. 4.4 Milk carton benches at the Expo Site

sterilized pack is made of paper, aluminum, and plastic. After being recycled and processed, cartons can be transformed into boards, which are durable, waterproof, and damp-proof. Each bench made of 856 milk cartons, has a capacity of 250 ml each. A great deal of milk cartons are thrown away everyday, if we recycle them, we can not only save resources, but also substantially improve the recycling rate of trash, and relieve the burden of environment.

4.3 Creative Furniture

Upon entering the London Case Pavilion, visitors could see the furniture and ornaments made from waste materials all over the building (Fig. 4.5). At the gate of the pavilion were a few desks and chairs scattered in the first hall, made of deserted gasoline barrels. Cut into different shapes and added into cloth art, these barrels seemed to be revitalized, giving out a feeling of warmth and peace. Inside the pavilion, there were three models made from disposable dishes and chopsticks, more or less with some defect, either having broken legs or arms, or stabbed through the

Fig. 4.5 Chairs and stools made of waste in London Case Pavilion

heart by trash. In delivering an aesthetic feeling to the tourists, they also confide to them the damage the environment has suffered from the throw-away dishware.

Inside the pavilion, a small meeting room attracted many tourists who hold their footsteps there to see the originality "stools" made from all kinds of waste. The stool rolled up from waste newspapers is solid and comfortable; the chair forged from waste iron pan with a cushion on it, bears great resemblance to a sofa; in addition, there are chairs pieced together from tap water pipes and stools made from waste liquid gas tanks.

Even the various kinds of droplights were "original home furnishing" made from different wastes like used drinking bottles. Chopsticks are edible, while plates are recyclable and degradable. As for the throw-away tableware that cannot be eaten, together with other food waste in the kitchen, are collected into the bio-energy stove, to generate electricity and heat (Fig. 4.6). As a matter of fact, all things in life can be converted into wealth if properly dealt with.

Fig. 4.6 Biodegradable tableware

4.4 Environmental-Friendly Paper

The hand tissue used in the toilet was especially developed and produced for the 2010 Expo by an enterprise from Shanghai. The raw materials for this kind of tissue all come from packing boxes of milk beverages, without cutting down even a single tree in the production process. As it is not involved in the normal bleaching process, this tissue may seem a little yellow, but it is safe and sanitary without any fluorescer, reducing the emissions of sewage and exhaust gas to a large extent. It was estimated that 2,000 tons of household tissue would be used during the World Expo. By using paper that is good for environment protection (Fig. 4.7), it can save 7,000 m^3 of wood and prevent 61,178 trees from being cut down.

4.5 Direct Drinking Water

The environmentally friendly and pollution-free drinking water system can provide drinkable water directly, which reduces the emission of a large amount of trash and accords with the low-carbon concept of World Expo 2010. The production of this drinkable water mainly adopted the high-tech processing technology of ultra filtration membrane, with key steps of "filtration by granular activated carbon, separation through ultra filtration membrane component and disinfection by ultraviolet tube." The treatment technique is low-carbon, green, and increasingly mature. "Compared with traditional filtration methods, this process avoids the possible chemical transformation; therefore it removes the potential harm to human health, and still it retains a considerable amount of minerals."

Fig. 4.7 Environmental-friendly paper at Expo site

4.6 Green Bicycles

The Danes brought their most proud bicycles to the Shanghai World Expo. Inside the Denmark Pavilion, a double-spiral ramp was especially designed, circling twice from the ground to reach a height of 12 m, and then spiraling back down to the ground. Visitors could choose to ride a bike or walk to tour around the pavilion.

Fig. 4.8 Bicycles in Denmark Pavilion

Fig. 4.9 Touring around Denmark Pavilion by bike

The bicycle is a commonly used means of transportation in Denmark, and in order to promote the idea of green travel, the Denmark Pavilion prepared about 300–400 bikes for the visitors (Figs. 4.8 and 4.9).

"The purpose of cycling around the Expo pavilion is to enable visitors to practice the concept of green travel by themselves," said the spokesman of Denmark Pavilion in an interview. And to build a pavilion unique like this is to tell people that they do not have to drive a car whenever they go out, and that riding a bike is not only more convenient, but also full of fun. If people go to work on bikes instead of driving, and suppose that the distance for driving is 5 km everyday, then by the end of a year, everyone can reduce the emission of carbon dioxide by 300 kg every year.

Chapter 5
Conclusions and Recommendations

In the Shanghai Expo, low-carbon concept was embodied everywhere from the construction of the Expo Site to the demonstrations of various technologies. All the participating countries' pavilions applied and showcased new concepts and the cutting edge green technologies in their field through exhibitions and by the architecture itself.

From the overall planning and operations of the Expo site, the practices in low-carbon technologies application were mainly reflected in the construction of the core buildings know as "One Axis and Four Pavilions" (The Expo Center, the Performance Center, the Theme Pavilions, the China National Pavilion and the Expo Axis) and the public service facilities of the site.

Buildings in "One Axis and Four Pavilions" combined up-to-date technologies of the relevant industries both at home and abroad in its architectural design, building energy efficiency, new energy application, and low-carbon idea. They included the rainwater recycling system and water source heat pump system applied in the Expo Axis; the Solar Photovoltaic wind power, river water source heat pump, ice Storage, natural lighting systems incorporated in the Expo Center; the large-scale application of Solar photovoltaic power generation in the China National Pavilion, and the Theme Pavilion. Besides, energy-saving building materials were extensively used in these buildings.

Public service facilities mainly include public transport, waste disposal, water treatment, and the overall planning and construction of public service establishments. Specifically, public transport brought together various types of new energy vehicles with multiple functions, such as supercapacitor buses, fuel cell vehicles, pure electric, and hybrid electric vehicles (HEV) manufactured by numerous enterprises. Low-carbon technologies were also applied in trash disposal, such as the pneumatic dustbin recycling system, electric garbage disposal cars, and sanitation trucks. Water treatment technology was incorporated into the overall planning and construction of the whole Expo site, as seen from the recycling of rainwater and sewage. Besides, low-carbon technology also contributed to the

UNIDO-ISEC and Jiangsu Modern Low-carbon Technolgy Research Institute (eds.),
The Research Report on Application of Low-carbon Technology in Expo 2010 Shanghai,
DOI 10.1007/978-3-662-44357-6_5

installation of cooling atomizer, solar-powered mosquito killing light, and meteorological observatory and LED lamps in public service places.

From the architectural and structural points of view, low-carbon concept was reflected in the infrastructure of all pavilions, from the new energy, architectural design, building energy-saving technology to the new building energy-saving materials. In this regard, the gap between developed countries and developing countries was obvious. Regardless of the proportion of the technology applications, the technological maturity, or the completeness of low-carbon industrial chain system, the developed countries were far ahead.

Based on the data obtained from the survey, conclusions are reached as follows:

Compared with new energies, energy-saving technologies are more likely to be popularized, with the popularizing rate of building energy-saving relatively high. The energy-saving technologies are directly reflected in building energy efficiency and consumption, and meanwhile they can meet the benefits of both the buyer and the seller. As for new energies promotion, it has to go through several phases such as technological maturity, policy guidance, market acceptance and test, therefore the promotion and popularization of new energies are comparatively more difficult.

The European countries and the United States still take the lead in the field of low-carbon technology, including technology, concept, popularizing rate of application, etc. But at the same time, quite a number of Chinese enterprises also demonstrated the powerful strength in their own fields.

To embed the low-carbon concept into everyday life is much more important than to popularize the technology. There is no doubt that the research and development of low-carbon technology is of great significance, but if the low-carbon concept of environmental protection and awareness is not deeply rooted in people's mind, such prospects as low-carbon economy and low-carbon life are nothing but an illusion. That is why most of the things on display in developed countries' pavilions were more about how to help people actually internalize the low-carbon concept.

To conclude, low-carbon is not a temporary idea, but an irresistible trend for social development, and it will inevitably become an ultimate choice.

Recommendations are made in the following aspects:

(1) **Solar Energy**

First, it is recommended that emphasis should be put on the development of solar concentrated photovoltaic (CPV) and solar thermal integration technology, which is, to integrate thermal utilization into photovoltaic power generation, much importance should be attached on amorphous silicon solar technology research and development. Second, vigorously carry out the research and development of China's solar thermal technology and solar thermal industry. Although the solar thermal industry in China has just started, its current speed of development is very rapid. It is recommended to make the complementary use of solar thermal power generation and other energy sources, to conduct research of distributed energy applications, for example, solar air-conditioning technology, solar desalination of seawater and so on.

(2) **Building Energy-Efficiency**

It is recommended that efforts should be put on steel structure buildings development based on the following reasons:

a. China faces excess steel capacity;
b. Steel structure can be recycled;
c. Steel structure produces less building waste and pollution at site assembling, avoid pollution caused by the concrete production, and meanwhile reduce the construction cycle significantly;
d. Development of steel structure will drive the development of a series of new industries and industry chain;
e. Steel structure building is the optimum type of construction to resist damage in earthquake-prone areas.

(3) **LED New Light Source**

According to statistics, the power consumption of lighting in China accounts for 12 % of the total amount, while the proportion in the developed countries is 20 %. LED light saves 90 % power consumption than traditional illumination lamp, so it is necessary to promote the LED lighting technology. It is recommended that the government should promulgate the relevant policies and tackle the key problems in support of the development of the LED industry.

(4) **Biomass Energy**

Strive to develop biomass energy technology and carry out the study of vehicle-mounted biomass based on human, livestock, poultry, domestic sludge, agricultural straws, and methane technology so as to make recycling utilization of wastes and minimize pollution caused by traffic emission.

(5) **Wind Power**

Our recommendation is to speed up the development of light turbines used in urban areas and promote their distributed applications, and support the research and development of the turbines with the independent intellectual property rights.

(6) **Geothermal Energy**

Greatly promote and apply ground/water source heat pump air-conditioners, use less electricity so as to reduce the operating cost and carbon emission. Put much effort on the R&D of various kinds of energy storage technologies.

(7) **Trash Collection**

Introduction of compulsory measures and market operation mechanism, and make classified garbage collection, to mitigate the difficulty in garbage disposal and therefore increase the waste recycling utilization.

(8) **Low-Carbon Transportation**

Conduct study on the renewable road surface material and utilization of other waste materials. Vigorously carry out study on road noise abatement technology. Take steps to improve the driver's driving skills and other quality.

(9) **Water Treatment**

It is recommended to increase rainwater and reclaim water collection and use; increase the use of wastewater treatment and reuse of gray water. Strive to tackle the problem of water purification. Under the premise of ensuring water quality, we inspire the promotion of direct drinking water in China. This will not only reduce energy consumption but also reduce carbon emissions. More importantly, increase the public awareness of current water resource shortage, so that more and more people will pay close attention to water conservation in their daily life.

Appendix A
Companies and Organizations in Application of New and Renewable Energy Technologies in the Expo

A.1 Solar Energy Companies and Organizations

(See No. 1)

No. 1

No.	Company name	Website	Pavilions involved
1.1	Beijing Micoe Solar Energy Co., Ltd.	http://www.micoe.com	USA Pavilion, Information, and Communication Pavilion
1.2	Delft University of Technology	http://www.tudelft.nl	Netherlands Pavilion
1.3	Fuji Electric Holding Co., Ltd	http://www.fujielectric.com.cn	Japan Pavilion
1.4	Himin Solar Co., LTD	http://www.himin.com	Private Enterprises Joint Pavilion, London's UBPA Case
1.5	Hunan Sun-zone Optoelectronics Energy Co., Ltd.	www.sunzone-energy.com	China Pavilion
1.6	Jiangsu Huayang Solar Energy Co., Ltd	http://www.huayangsolar.com	Private Enterprises Joint Pavilion
1.7	Ecole Polytechnique Federale de LAUSANNE.	http://www.epfl.ch	Swiss Pavilion
1.8	Ningbo KangLiTe Electronic Co., LTD	http://www.klt123.com.cn	Private Enterprises Joint Pavilion
1.9	Nanjing Nanzhou New Energy R&D Corp.	http://nz-newenergy.com	UN Pavilion

(continued)

UNIDO-ISEC and Jiangsu Modern Low-carbon Technolgy Research Institute (eds.),
The Research Report on Application of Low-carbon Technology in Expo 2010 Shanghai,
DOI 10.1007/978-3-662-44357-6

(continued)

No. 1

No.	Company name	Website	Pavilions involved
1.10	Norway Malthe Winje AS	http://www.malthe-winje.no	Norway Pavilion
1.11	Suntech-power Holding Co. Ltd	http://suntech-power.com	China Pavilion, Theme Pavilion, Expo Cultural Center, Expo Center, Expo axis, Macau Pavilion, Solar cruise
1.12	Shanghai Fanxing Lighting Co., LTD.	http://www.shfxzm.com	Expo Site
1.13	Shanghai Guosheng (Group) Co., Ltd.	http://www.sh-gsg.com	Shanghai Corporate Joint Pavilion
1.14	Shanghai Jiulong Power Group	http://www.jldl.com.cn	Pavilion of Future
1.15	Shanghai Jiaotong University	http://www.sjtu.edu.cn	State Grid Pavilion
1.16	Shanghai Solar Energy Technology Co., Ltd.	http://www.ssec-solar.com	Expo Center
1.17	Shanghai Solar Energy Research Center	http://www.solarcell.net.cn	Expo Center
1.18	Trinasolar.Energy Co., Ltd.	http://www.trinasolar.com	Belgium—EU Pavilion
1.19	Wacker Co., Ltd.	http://www.wacker.com	Hamburg House
1.20	China Guodian Corporation	http://www.cgdc.com.cn	Finland Pavilion

A.1.1 Detailed Information About the Companies and the Products

(See No. 1.1, 1.2, 1.3, 1.4, 1.5, 1.6, 1.7, 1.8, 1.9, 1.10, 1.11, 1.12, 1.13, 1.14, 1.15, 1.16, 1.17, 1.18, 1.19 and 1.20)

No. 1.1

Company name	Beijing Micoe Solar Energy Co., Ltd.	Application Pavilion	USA Pavilion, Information and Communication Pavilion
Website	http://www.micoe.com		
Key products and services	Beijing Micoe Solar Energy Co., Ltd. is a leading enterprise in China's solar industry, especially in solar photo-thermal utilization. As a high-tech enterprise, Micoe is specialized in BIPV (Building Integrated Photovoltaics), with the business covering solar water heater, product development, manufacture, and application. It was the first to achieve the output of over 1 million units in solar industry in 2009. For the past 10 years it has promoted the solar energy application with an area of 12 million m^2. Micoe has been promoting the industrialized development through technological innovation, which is the only one to achieve the whole industrial chain in research and manufacture from pipes to water heaters. Micoe has four different technologies such as all glass evacuated solar collector tubes, flat plate, glass, and metal solar collector tubes, and can manufacture the four kinds of products, providing more options and support for different market demands		
	Micoe provided the USA Pavilion, Information and Communication Pavilion with the cutting-edge full automatic water heating system leading the industry worldwide. The 24 h continuous water heating solutions provided for the two pavilions is intelligent and low-carbon, of which the project in the USA Pavilion is the biggest water heating system engineering in the Expo		
	Advantage in Innovative: highly mixed rare metal aerospace material coating technology, double anti reflection coating technology, unstable state heat efficiency experiment method, green freon-free foaming production line, sophisticated SWT testing instruments, etc.		
Patent information	Over 20 items of independent intellectual properties, undertaken a number of projects under the State 863 Plan		

No. 1.2

Company name	Delft University of Technology	Application pavilion	Netherlands Pavilion
Website	http://www.tudelft.nl		
Key products and services	Delft University of Technology is one of the world's leading Universities of technology, also known as the MIT of Europe. The high speed solar car developed by Delft University of Technology is single solar-powered car which can speed up to 145 km/h. It has won four consecutive world champions in the World Solar Challenge race held in Australia. Nuna5, covered by 6 m^2 of solar panels, also with a car battery to store excess electricity. The car can still run additional distance when there is insufficient sunlight. When necessary, it can also accelerate the speed to improve its competitiveness in the race		
Patent information	Self-independent patent		

No. 1.3

Company name	Fuji Electric Holding Co., Ltd	Application Pavilion	Japan Pavilion
Website	http://www.fujielectric.com.cn		
Key products and services	Fuji Electric Co., Ltd., founded in 1923, is the world's leading multinational corporations and Japan's largest integrated manufacturers of electronic products		
	Fuji Electric Holding Co., Ltd. mainly engages in the following five areas:		
	Drive: drive control, uninterruptible power systems, motor/ applications; *Automation*: radiation monitoring and management systems, measurement devices, power control systems, power metering, clean energy (solar, wind, micro-irrigation, power generation systems such as fuel cells);		
	Industrial equipment engineering: facilities monitoring and control systems, transportation equipment and systems;		
	Power generation equipment engineering: energy generation, geothermal power, hydropower;		
	Semiconductors: Fuji Electric Holding Co., Ltd. provides green energy, power grid, environmental industry, transport, and other low-carbon energy solutions to the world through its own technological advantages		
	Its application in the Expo is by using the transparent oxide semiconductor to convert the solar energy into power		
Patent information	Independent patent		

No. 1.4

Company name	Himin Solar Co., Ltd.	Application pavilion	Private Enterprises Joint Pavilion, London's UBPA Case
Website	http://www.himin.com		
Key products and services	Himin Solar sells 3 million m^2 of solar heaters annually, with the products ranging from solar water heaters (home solutions), split solar water heating systems, solar power generation, solar air-conditioning, to solar swimming pool and seawater desalination, etc.		
	Solar photoelectricity scenery is the new area for solar PV terminal use, including Solar photoelectricity sculpture, solar landscape lamp, gazebo, BIPV PV shadow gallery, etc. It combine solar photoelectricity, scenery art as well as lighting together to realize the perfect combination of sculpture scenery and solar PV, and it acted as normal scenery at the daytime but shining at night		
	Application in the Expo:		
	Solar photoelectricity sculpture, solar photoelectricity pavilion, BIPV PV shading gallery		
	Characteristics:		
	1. High IPCE and reliability.		
	2. electric energy conservation		
	3. No polluted emission while working		
Patent information	With 215 national patents, undertaking or participating in 22 national projects under the State 863 Plan, the State Scientific and Technique Plans, National Torch Plan		

No. 1.5

Company name	Hunan Sun-zone Optoelectronics Energy Co., Ltd.	Pavilion for application	China Pavilion
Website	http://www.sunzone-energy.com		
Key products and services	Hunan Sunzone Optoelectronics Energy Co., Ltd. was established in August 2008. The company initially invested 580 million yuan, and got 155 acres of land acquisition in building 200 MW vertically integrated solar cell industry chain, including single and polycrystalline silicon solar cell production, solar modules packaging, and solar application system design and installation. Solar cells: efficient color polycrystalline silicon solar cells; efficient color single silicon solar cells; monocrystalline silicon solar cells; polycrystalline silicon solar cells. Solar battery modules: color double glass solar modules; crystalline silicon solar cell modules; monocrystalline silicon solar modules. Solar PV systems: efficient solar power system		
	The 250 kW efficient color solar power generation and grid-connecting system installed on the roof of the China Pavilion consists of 2,736 high-color (red, green, blue) solar cell modules, and was formally put into operation on February 10, 2010, which is for the first time that the efficient solar power system was applied in the building		
Patent	Patent technologies: solar photovoltaic systems self-cleaning, cooling system; a colored solar cells; a colored solar cell preparation method; a full solar tracker; a crystalline silicon solar cells on the surface of the “dead layer” treatment process; a surface coating solar modules glass and its preparation method		

No. 1.6

Company name	Jiangsu Huayang Solar Co. Ltd.	Application Pavilion	Private Enterprises Joint Pavilion
Website	http://www.huayangsolar.com		
Key products and services	The company specializes in new energy development and application as well as green hot water solutions. It has got more than 170 items own proprietary technologies, including common type of solar water heaters, all-day intelligent solar water heaters, balcony hanging solar water heater, villa solar water heater, and building integrated solar water heaters, large area hot water system for groups, home or group air heat pump water heaters, photovoltaic and other optoelectronic products. The products sell to dozens of countries and regions, including the United States, Russia, Germany, Italy, Singapore, South Korea, Taiwan, etc.		
Patent	Self-independent patent		

No. 1.7

Company name	Ecole Polytechnique Federale de LAUSANNE	Application pavilions	Swiss Pavilion
Website	http://www.epfl.ch		
Key products and services	EPFL was founded in 1853. It is an international university, with the students coming from 107 different countries and regions. EPFL consists of seven colleges, namely the basic science, Engineering, environmental, construction and civil engineering, computer communications, College of Life Science, Technology School of Management and Humanities. Swiss Museum of solar cell technology provided by the EPFL		
	Outside the Swiss National Pavilion, the 17-m-high translucent curtain is trimmed with 10,000 red solar cells called "Little Sun", which is disc-sized red photosensitive device. With the interaction of solar cells, light sensors, and LED, it absorbs the light energy and converts it to electricity and shine. The window glass is a "sandwich" structure solar cell, with the secret of the photoelectric conversion hidden in the nano "filling" between the hollow glass. This "dye sensitized" solar cell was first invented by the scientists of Swiss Federal Institute of Technology in Lausanne (EPFL), inspired by the photosynthesis of the nature. It is like the man-made "chloroplast" on the plant which costs only 1/10 of the silicon-based solar cells		
Patent	Self-independent patent		

No. 1.8

Company name	Ningbo Kanglite Electronic Co. Ltd.	Application pavilion	Private enterprises Joint Pavilion
Website	http://www.klt123.com.cn		
Key products and services	Ningbo Kanglite Electronic Co. Ltd. is a manufacturer of solar cells, specializing in the development and manufacture of the solar cell plates. The products include solar power generating unit, solar cells and cell plates, applied in the lawn lamps, yard lamps and toys, etc.		
Patent	Self-independent patent		

No. 1.9

Company name	Naning Nanzhou New Energy Research & Development Corp.	Application pavilion	UN Pavilion
Website	http://nz-newenergy.com		
Key products and services	Nanjing Nanzhou New Energy Research & Development Corp. develops the building integrated solar water heaters, with its new energy and modern architecture, and cultural integration. The new type "Qin Brick" photovoltaic system on wall space has the increased conversion rate of 4.2132 times compared with the world's popular wall silicon photovoltaic cells. The amount of silicon cells per unit lighting area in this new optical Qin brick is only 0.2373 times of that in the existing optical building wall. It is suitable for a the roof of a variety of urban and rural construction and integration of solar thermal systems, and is also be applied in earthquake relief, the military, geological, temporary buildings, and old building reconstruction		
	The product was applied in the waiting area of the United Nations Pavilion. The area is 40 m long, 5 m wide, and 3.8 m high, with an area of 232 m^2. A total of 150 pieces of "New Tiles" were laid, with the total installed capacity of 15.12 kW		
Patent	New Tiles roof system patent, New Qin Brick wall system patent		

No. 1.10

Company name	Malthe Winje AS	Application pavilion	Norway Pavilion
Website	http://www.malthe-winje.no		
Key products and services	Malthe Winje AS was founded in 1922. With more than 85 years history the Group is a holding company with the scope of activities as:		
	Electro technical products Low Voltage		
	Electro technical products High Voltage		
	Automation and PLC systems		
	Smart House Solution		
	Automation/Remote Control projects within:		
	Water and Sewage Treatment		
	Power stations and distribution (High Voltage)		
	HVAC (building automation and marine automation		
	Traffic Control systems/equipment		
	Real estate development		
	Norway Pavilion showcased a water purifying plant powered by solar energy, which can change the polluted rain into clean drinking water. It can eliminate all pollutants from water including germs, virus, and parasites. The process is not environmentally hazardous. This system can be transport in standard containers		
Patent information	Self-independent patent		

No. 1.11

Company name	Suntech Power Holding Co., Ltd.	Application Pavilions	Macau pavilion, Expo Axis, Expo Center, Expo Cultural Center, Theme Pavilion, China Pavilion, Solar cruise
Website	http://suntech-power.com		
Key products and services	Suntech Power (Wuxi) has its unique silicon wafer texture and cleaning process, cutting-edge PECVD film-forming technology and designs on silver positive electrode and silver aluminum back electrode. 80 % of the company's business lies in the US, Europe, the Middle East and Southeast Asia. High-performance features and superior reliability make Suntech products the best choice for customers around the world. In China the products have been extensively applied in the fields such as communications, broadcasting, transportation, petroleum, lighting, etc. The company is committed to the localization of the photovoltaic industrial chain as well as large-scale development of the industry in China, and is making great effort in speeding up the R&D in the industrialization of the second generation Poly-Si thin film solar cell, to keep a leading position in the industry		
	Suntech can supply the most complete range of photovoltaic modules in the industry such as See Thru, Light Thru, Semi-Shade, Just Roof and other series of products. Its core product is the standard photovoltaic module which enjoys good reputation, made of glass shell, and TPT baseboard. In fixation, it needs additional fixed point on the building structure such as supports on the rooftop. Suntech's standard PV module has been used widely around the world such as San Francisco International Airport, Greater Murcia's civil projects. Suntech's See Thru, Light Thru, Semi-Shade, Just Roof are all products of BIPV. See Thru module uses the A-Si (noncrystalline silicon) thin film PV technology which allowed transmittance of about 1, 5 or 10 %. It can be used in curtain wall, skylight, ceiling, porch, and vertical or slant polished surface. Monocrystalline silicon or polysilicon solar battery are added between the two glasses of Light Thru module which is packed with EVA or PVB and pressed between two low-iron toughened glasses or between glass (positive side) and Tedlar (back side). It can be used in ceiling, porch, skylight, and vertical curtain wall		
Patents	Suntech's independently developed noncrystalline silicon patent technology, Pluto cells		

No. 1.12

Company name	Shanghai Fanxing Lighting Co., Ltd.	Application pavilion	Expo Site
Website	http://www.shfxzm.com		
Key products and services	Fanxing Lighting is a research-based manufacturer of the energy saving and environmental protection lighting products. The products for energy saving include the three primary colors electronic energy saving lamps and high quality electronic ballasts; for environmental protection, the solar energy lighting, and solar energy black light pest control devices, which were widely used in the Shanghai World Expo		
	Solar black light lamp: by use of solar photovoltaic technology to convert solar energy into electric energy; with the new broadband efficient black lamp as lure insect light source, together with the intelligent control system to kill the pests touching the high-pressure metal network		
	Taking advantage of the pest's phototaxis, many kinds of pests can be killed with this device. A certain amount of trapping of the pests that are the main threaten of the large green gardens is seen. About 150 such devices were placed in the Pudong site of the Expo and 70 in the Puxi site		
Patent	Self-independent patent		

No. 1.13

Company name	Shanghai Guosheng (Group) Co., Ltd.	Application pavilion	Shanghai Enterprise Joint Pavilion
Website	http://www.sh-gsg.com		
Key products and services	Shanghai Guosheng (Group) Co., Ltd. is approved the state-owned large-scale investment holding and capital operating company approved by Shanghai Municipal People's Government. It provided solar hot water power generation for the Shanghai Enterprise Joint Pavilion		
	Solar hot water power generation: 1 ton hot water in 90 °C can generate 3.5 kW of electricity, the solar power generation device in the night can still work using the residual heat of water, the generated electricity can meet architectural exhibitions and daily electricity needs. On the roof is 2,200 m^2 of solar heat collecting screen covered with pipes which are full of water. In the case of absorption of solar energy, the pipeline can produce 95 °C hot water. With the low-temperature hot water power generating technology, the output power reaches over 200 kW		
Patent	Self-independent patent		

No. 1.14

Company name	Shanghai Jiulong Power Group	Application pavilion	Pavilion of Future
Website	http://www.jldl.com.cn		
Key products and services	Shanghai Jiulong Power (Group) Co., Ltd. was established in 1995, the first group holding company in Shanghai power system. Jiulong Power focuses in service-oriented business in power engineering at home and abroad, including design and planning of transmission and distribution, equipment manufacturing, power facilities installation, capital operation, ancillary services, etc.		
	Solar PV electric products applied in Pavilion of Future features:		
	1. High IPCE and reliability		
	2. Advanced diffusion technique, Guaranteed uniform conversion efficiency in the pieces		
	3. The tracking solar system improved the application of solar energy		
	4. Amorphous silicon battery pack has the merits of better scattered light and lowcost		
Patent information	Independent patent		

No. 1.15

Company name	Shanghai Jiaotong University	Application pavilion	State Grid Pavilion
Website	http://www.sjtu.edu.cn		
Key products and services	Shanghai Jiaotong University is one of the universities with the longest history, and national key university directly under the Ministry of Education. It provided the following technical support for the Shanghai Expo:		
	a. The application of air conditioning technique driven by transformer's residual heat at State Grid Pavilion: The cooling waste heat of 110 kv transformer can drive 2 adsorption refrigeration packs whose service conditions are: when hot-water temperature is 60–65° C, refrigerating capacity reaches more than 10 kW, undertaking part of the cooling load of the substation's air-condition		
	b. The application of natural draft techniques enhanced by solar power at State Grid Pavilion: 150 m^2 vacuum pipes solar collector integrated with the building roofs. Heating the atrium by solar collector through steel radiator		
Patent	Independent patent		

No. 1.16

Company name	Shanghai Solar Energy Technology Co., Ltd	Application pavilion	Expo Center
Website	http://www.ssec-solar.com		
Key products and services	Shanghai Solar Energy Technology Co., Ltd. (SSEC) is a stock company jointly established by Shanghai Aerospace Automobile Electromechanical Co., Ltd., Shanghai Shenergy New Energy Investment Co., Ltd., Shanghai Institute of Space Power. As a high-tech enterprise SSEC specializes in the power system development and production, including ground solar cell systems. The products have a wide range of application areas such as the solar power plant, wind/solar hybrid power plants, solar PV pumps, marine aids to navigation, radio and television, mobile base stations, and grid connecting and solar users systems		
	Main products: Solar energy cells, single crystal silicon solar cells and modules, Polysilicon solar cells and modules, PV integration, PV independent system, PV grid connecting system, and BIPV		
Patent	Independent patent		

No. 1.17

Company name	Shanghai Solar Energy Research Center	Application pavilion	Expo Center
Website	http://www.solarcell.net.cn		
Key products and services	Shanghai Solar Energy Research Center is a municipal research and development center established with the approval by Shanghai Science and Technology Commission in 2005, specializing in research, development, and application of solar photovoltaic technology. The business covers the technology consulting and development within the field of solar energy, and related product development, product testing, related development of the testing and production equipment, technology consulting, and services		
	The Center has developed the special solar PV modules, such as PV sun shading, PV glass curtain wall, Two-side PV glass modules, efficient flexible thin film solar cell, CIGS thin film cells, high efficiency PV concentrator cells, single- and double-shaft concentrator cells tracing systems, infra-red testing series devices, spot testing devices, hail testing unit, mechanical load, high-low temperature cycles testing devices, and other products. These self-developed products prove good using effect and have great market share and influence		
Patent	Independent patent		

No. 1.18

Company name	Trina solar Energy Co., Ltd	Application pavilion	Belgium—EU pavilion
Website	http://www.trinasolar.com		
	Trina solar Limited (TSL) is a global manufacturer of crystalline silicon solar modules and has a long history since it was founded in 1997. In 2006 it went public in New York Stock Exchange (NYSE). Based on its strong vertically integrated business model, Trina develops and manufactures the ingot, wafer, cells, and solar modules. In 2010, TSL was regarded as the industrial pioneer in World Economic Forum in Davos. It ranks second in terms of the sustainable index of the 2010 Version global PV industry published by PRTM		
	In 2010, the state PV science and technology key lab was approved to be built in TSL, which lays a solid foundation for pursuing the outstanding quality. Today TSL develops into the leader in the solar industry and maintain one of the lowest cost structure in the industry. The company also pays much concern on investment in safety, environment, and health		
	TSL is headquartered in Switzerland and USA, and has offices in Seoul, Tokyo, Madrid, Milan, Beijing, Shanghai, Sidney. The PV modules products are exported to more than 20 countries and regions in Europe, USA, Asia-pacific region		
	TSL provided the 160 kW solar power generation system for the Belgium-UN pavilion, covering an area of 1,200 m^2, with annual output reaching 116,000 kWh, and achieve the substantial reduction of carbon emission		
Patent	Independent patent, the number of items for patents, and inventions ranks the first in China's PV industry		

No. 1.19

Company name	Wacker Co., Ltd	Application pavilion	Hamburg House
Website	http://www.wacker.com		
Key products and services	Wacker Chemical is a world famous diversified company, headquartered in Munich, Germany		
	Its main business involves developing and manufacturing organic silicon, polymer, fine chemical, poly-Si and hyperpure silicon wafer		
	Core technology: (a) using ketene (CA) to produce Aceticanhydride; (b) patent of suspension method producing PVC-U; (c) Polishing ultra purified silicon chip		
	Wacker is engaged in the photovoltaic solar thermal energy and wind power generation in the renewable energy field		
	The 450 m^2 PV power generating devices installed in the Hamburg House supplied 80 % of its electricity consumption needs. Wacker also provided a series of solutions for the Hamburg House such as silicon protection agent for concrete treatment, organic silicon seal rubber, and tile adhesive containing high performance dry mortar and seam filling mortar. These materials, together with other building materials built a high-effect Hamburg House		
Patent	Independent patent		

No. 1.20

Company name	China Guodian Corp.	Application pavilion	Finnish Pavilion
Website	http://www.cgdc.com.cn		
	China Guodian Corp is a comprehensive electricity group founded in 2002, approved by the state council. It is engaged in the development, investment, construction, operation and management of the power source, organizing production and marketing of the electricity (heat); investing in any electricity-related business such as coal, new energy, transportation, high-tech, environmental protection, technical service, information consulting, funding at home and abroad, international cooperation, etc.		
	Application in Finland pavilion: supply electricity for the lighting of the pavilion by converting the solar energy into electricity with solar cells. A total of 162 monocrystalline ingots 180 W solar PV modules were used, and 30 kW solar grid connecting projects		
Patent	Independent patent		

A.2 Wind Power Companies and Organizations

(See No. 2, 2.1, 2.2, 2.3, 2.4, 2.5, 2.6, 2.7, 2.8, 2.9, 2.10 and 2.11)

No. 2

No.	Company name	Website	Pavilions
2.1	Bosch Rexroth, China	http://www.boschrexroth.com.cn	Pavilion of Urban Planet
2.2	EDP group	http://www.edp.pt	Portugal Pavilion
2.3	EWT international	http://www.ewtinternational.com	Netherlands Pavilion
2.4	Guodian United Power Technology Co., Ltd.	http://www.gdupc.com.cn	MeteoWorld Pavilion
2.5	Sinovel Wind Group Co., Ltd.	http://www.sinovel.com	Model demonstration in Pavilion of Future
2.6	Maanshan Fangyuan Slewing Ring Co., Ltd.	http://www.masfy.com	China Pavilion
2.7	Nantong Zi lang Wind Turbine Co., Ltd.	http://www.ntzlfd.com	Pavilion of Future
2.8	Panasonic Co., Ltd.	http://panasonic.co.jp	Japan Pavilion
2.9	Shanghai Ture Enclosure Electrical Co., Ltd.	http://www.turebox.com	Pavilion of Future
2.10	Shanghai Forevoo Windpower Technology Co., Ltd.	http://www.forevoo.com	World Expo ferry
2.11	Shanghai Aeolus Wind power Technology Co., Ltd.	http://www.sawt.com.cn	India Pavilion, World Expo Mobile Base Station

No. 2.1

Company name	Bosch Rexroth, China		Application pavilions
Pavilion of Urban Planet			
Company website	http://www.boschrexroth.com.cn		
Key products and services	Bosch Rexroth is one of the world's leading specialists in the field of drive and control technologies. Under the brand name of Rexroth the company supplies customers with tailored solutions for quality electric control, hydraulics, pneumatics, and electric drive and controls components and systems		
	Main products MOBILEX GFB variable propeller driving device, quiet and reliable REDULUS GPV electric generator gear box, compact REDULUS GPV-D differential gear box: Mega-Watt MOBILEX GFB yawing driven device, driving chain brake, and yawing brake		
Patent	Independent patent		
Products and models	MOBILEX GFB variable propeller driving device		
MOBILEX GFB variable propeller driving device			
Technical characteristics			
Output torque : 3–28 kNm			
Gear ratio: 60–200			
Two-stage or three-stage epicyclical gear box of compact type and space intensive type			
Optional bevel gear system			
Firm and durable alkaline-friction bearing			
Provided attachment of motor			
Simplified installation			
Easy to renewal lubricant oil			
Low operation noise			
Variable propeller system			

Technical characteristics of motor-driven paddle adjustment:	*Technical characteristics of electricity-hydraulic driving system:*
Alterable power range	Accurate controllability
Maintenance free AC-motors	High degree of integration
	Compact design
UPS of double-layer capacitors	Reliable Fail-safe Function
Driving controller of feedback function	Integrated security function
	Simple commissioning operation
Emergency operation function with no need for driving controller	Longer service life
	Highly guaranteed environment compatibility
Every spindle is equipped with independent energy management system	Flexible arrangement
On-line system diagnostic function	All versatile analog and digital interfaces

No. 2.2

Company name	EDP Group	Application pavilions	Portugal Pavilion
Company website	http://www.edp.pt		
Key products and services	EDP Group is a influential electric equipment provider in Europe, and one of Portugal's largest business groups and Portugal's largest energy company as well as one of the leading electricity operators in Europe. EDP Group was established in 1976 and became Portugal's largest private company in 1994. EDP Group's business mainly contains three main divisions: Electric Power division, Service Support division and International Division. Electric Power division provides power generation equipment and power transmission equipment, and has a unique advantage in power generation equipment with the renewable energy. Service Support division operates Portugal's largest private telecommunications network. International division is responsible for international market development and international cooperation, with a rich experience in international cooperation, with business in Latin America, Africa and Macao, China. EDP's main business is electricity, gas, fuel, with electricity installation capacity of 18,418 MW, power generating capacity of 48,903 GW/h, power distribution capability 166,084 GW/h; gas transmission capacity of 52,338 MW/h		
	EDP has also extends its business into water supply, gas supply, engineering and real estate, of which power generation, distribution and trade has spread to Spain, USA, Brazil, and Macao. The newly established Macao-based EDP Energy Solutions Asia focuses on exploring China's clean energy market, with its operation center in Beijing		
Patent	Independent patent		

No. 2.3

Company name	EWT	Application pavilions	Netherlands Pavilion
Company website	http://www.ewtinternational.com		
	EWT is a new wind energy company in Netherlands. At the end of 2003, EWT acquired the intellectual property rights of Lagerwey, the wind power pioneer brand (the company had produced 1,200 sets of various kinds of turbines). Currently, EWT's 200 sets 900 kW pitch speed direct-drive turbines have been used around the world, and becomes the world's leading direct-drive wind turbine suppliers with its self-developed direct-drive turbines		
	The features of the turbines:		
	Adoption of multi-stage air-cooled synchronous motors in EWT production, all-around back to back IGBT-type AC-DC-AC frequency control devices and the use of the independent axis and engine rotor impeller unit, replacing the traditional high-speed gear drive, greatly reducing the number of parts, proving the cutting-edge level in the development and manufacture of the direct-drive worldwide		
	Main products: DW 90—2.0 MW, DW 54—750/900 kW		
	Technology: direct drive technology, design of vane, DMS control system		
	Light component, high reliability, less maintenance, and high-efficiency		

(continued)

(continued)

Patents	Direct drive technology		
Products and models	DW 54—750/900 kW		
Description of the product	Wind-driven generators of 750 kW and the 900 kW uses advanced direct-driven technology, featuring:		
	Availability optimized to its maximum		
	Maintenance procedures simplified and maintenance time reduced substantially		
	Full variable speed		
	Full power back-to-back power converter		
	Less noise production than gear-box based turbines		
Model	DW 54/52	Adopted standard	IEC 61400 wind class III A (DW 54)
			IEC 61400 wind class II A (DW 52)
Cut in wind speed	2.5 m/s	Drive system	The air-filled synchronous generators full Power converters designed by EWT,
			IGBT'S AC & DC control Model
Rated power	750/ 900 kW	Control system	Bachmann PLC control system adopts TCP/IP interaction and DMS 2.0* SCADA system
Safe wind speed	59.5 m/s	Safe system	Service Brake System consists of three independent brakes and three independent emergency batteries
Rotor side variable frequency drive	12–28 rpm	Height of tower	50/75 m
Cut-out wind speed	25 m/s	Rated wind speed	13 m/s
Diameter of rotor	54.0/ 51.5 m	Output power control	Frequency control of motor speed

No. 2.4

Company name	Guodian United Power Technology Co., Ltd.	Application pavilions	MeteoWorld Pavilion
Company website	http://www.gdupc.com.cn		
Key products and services	Guodian United Power is headquartered in Beijing, with five wholly owned subsidiaries, three holding companies. The main business is on the large wind turbine design, manufacturing, sales, and service. The "State Key Laboratory for wind power equipment and system" as the first-class research and development platform in wind power technology field provides a powerful technical support for the United Power. Through innovation, company formed the 1.5 MW room temperature and low temperature series turbines, double-fed self-developed 3 MW offshore wind now has a small production capacity, 5 MW or more large Wind turbines with independent intellectual property rights of are in development		
	Research and Development:		
	The R&D products include those suitable for wind generating set under unusual climatic conditions such as on the sea, the tidal zone, the high elevation, the low wind speed, cold wind, no sandstorm, and so on		
	Wind power forecast and remote control technology, green manufacturing technology about weight reduction of wind turbine, repair and life extension of the essential part, vane material recycling, etc.		
	50 Hz/60 Hz 1.5 MW variable pitch and permanent frequency double-fed wind turbines		
	High elevation, low wind speed, big blade, 1.5 MW type		
	3 MW wind turbines on the sea		
	Electro-hydraulic/electro-magnetic coupling synchronization wind turbines		
Patents	Independent patent		

No. 2.5

Company name	Sinovel Wind Group Co., Ltd.	Application pavilion	Model exhibition in Pavilion of Future
Website	http://www.sinovel.com		
Key products and services	Sinovel Wind Group Co., Ltd. is China's first specialized high-tech enterprise that has independently developed, designed, manufactured, and marketed large-scale onshore, offshore and intertidal series of wind turbines adaptable to a global variety of wind resources and environmental conditions. In 2008, with the newly installed wind power capacity of 1,403 MW, Sinovel ranked No. 1 in China and No. 7 in the world; in 2009, Sinovel's newly added installed wind power capacity reached 3,510 MW, ranking No. 1 in China and No. 3 in the world; in 2010, Sinovel added installed wind power capacity of 4,386 MW, ranking No. 1 in China and No. 2 in the world		
	The main products of the company are SL3000 & SL1500 series of wind turbines. SL1500 series of wind turbines are China's first independently developed wind turbines adaptable to all wind zones and variable environmental conditions		

(continued)

(continued)

Patent	The products have fully independent intellectual property rights, with the cutting-edge technology these worldwide mainstream grid-friendly wind turbines were developed independently by Sinovel			
Product model	SL3000 series Wind turbines			
Parameters of wind generating set series of SL3000				
Product description	Independently developed by Sinovel with fully independent intellectual property right, the SL3000 series of wind turbines feature the mainstream grid-friendly wind turbines of both onshore and offshore. With the rated power of 3,000 kW, the rotor diameter of 90/100/105/113 m and the hub height of 80/90/100/110 m, this series have employed multiple world-leading patented technologies, have the low-voltage ride-through capability, and are adaptable to all wind zones, environmental conditions and grid codes and requirements in the world. In November 2009, this series won gold award of the CIIF 2009 (China International Industry Fair 2009). The first set of wind turbines was used in Shanghai Donghai Bridge Offshore Wind Farm, the first national offshore wind demonstration project in China			
Type	SL3000/90	SL3000/100	SL3000/105	SL3000/113
Wind area types	IEC IA	IEC IIA	IEC IIA	IEC IIIA
Rated power (kW)	3,000	3,000	3,000	3,000
Cut in wind speed (m/s)	3			
Cut out wind speed (m/s)	25			
Rated wind speed (m/s)	13	12.5	12	11.5
Survival winds (m/s)	70	59.5	59.5	52.5
Operation environment temperature (°C)	Normal temperature type: −10 to +45, low temperature type: −30 to +45			
Survival environment temperature (°C)	Normal temperature type: −20 to +50, low temperature type: −45 to +50			
Impeller diameter (m)	90	100	105	113
Swept area (m^2)	6,519	7,962	8,443	10,039
Blade number	3			
Gearbox structural style	2-stages planetary-gear and single stage parallel-axes gear			
Generator type	Doubly-fed asynchronous induction motor, water cooling method			
Rated output voltage of generator (V)	690			
Generator frequency (Hz)	50Hz/60Hz			
Generator rated rotational speed/range (rpm)	1,200/600–1,400 (50Hz)			
	1,440/800–1,600 (60Hz)			
Generator power factors	Capacitive 0.9 inductive 0.9			
variable pitch-powed controlling system	Actuating motor			
Yawing system type	Active			
Yawing driving controlled	Multilevel planetary-gear drived by frequency conversed asynchronous motor			

(continued)

(continued)

Braking system of air brake	Variable pitch controlled by impeller		
Braking system of mechanical brake	Hydraulic disc brake		
Controlling system	PLC and remote monitor		
Tower type	Pyramidal tower of steel structure		
Height of tower hub (m)	80/90	80/90/100/110	90/100/110
Remarks	Reference company's websites of other product types		

No. 2.6

Company name	Maanshan Fangyuan Slewing Ring Co., Ltd.	Application pavilions	China Pavilion
Company website	http://www.masfy.com		
Key products and services	One of the largest manufacturers in China specializing in the development, design, and manufacturing of the slewing bearings. The first public company in the industry		
	Main products:		
	No teeth Single-row slewing bearing, inner teeth three-row of column slewing bearing, outer teeth Single-row ball slewing bearing, inner teeth single-row ball slewing bearing, Large-scale single-row ball outer teeth slewing bearing, 1.5 MW Wind yaw bearing and plasma bearing, no teeth single-row ball slewing bearing, Hafu slewing bearing, Double-row ball slewing bearing, single—row ball helical slewing bearing		
	Single-row four-point contact ball slewing bearing include: Q series, HS series, 01 series, and L series. Cross-roller slewing bearing include: J series, 11 series		
	Double-row different diameter ball slewing bearing: 02 series		
	Double-row ball and three-row column slewing bearing: S series, 13 series		
	Applications: Wind power generation, excavators, automobile cranes, tower cranes, Hydraulic aerial cages, crawler cranes, port cranes, carriers, Ferris wheel, radar		
Patents	Patent products: Hafu slewing bearing, Large diameter slewing bearing measuring rods comparison equipment, filling machine used double rotary slewing bearing, double-row ball slewing bearing, the heavy trailer used double L shape slewing bearing, Worm drive slewing bearing, replaceable gear slewing bearing, ship unloader used four-point contact wire-race slewing bearing, outer teeth drive hafu slewing bearing, Medical CT machine used wire-race slewing bearing, slewing bearing choked sealing equipment, heavy load large slewing bearing		

No. 2.7

Company name	Jiangsu Nantong Zi lang Wind turbine Co., Ltd.		Application pavilions	Pavilion of Future
Company website	http://www.ntzlfd.com			
Key products and services	Nantong Zi lang was founded in 2001, specializing in production of 300 W–30 kW series Zi lang wind turbines and also engaged in the research, production and marketing of new energy products such as controller, inverters, etc. The company has passed the CE authentication			
Patent	Independent patent			
Products and models (W)	200, 300, 500, 1,000			
Parameters of 200 W, 300 W, 500 W, 1,000 W				
Rated power (W)	200	300	500	1,000
Maximum output power (W)	280	400	700	1,300
Rated rotating rate (rpm)	450	400	400	400
Rated wind speed (m/s)	6	7	8	9
Starting wind speed (m/s)	3	3	3	3
Working wind speed (m/s)				
Working temperature (°C)	−40 to 60			
Output voltage (v)	12/24	24	28	48
Wind turbine diameter (m)	2.1	2.5	2.5	2.7
Blade material	FRP			
Blade number	3			
Mode of speed control	Automatically deflecting			
Suggested battery capacity	2 12v100AH	2	12v150AH	2
2v150AH	4 12v150AH			
Output control system	Controller, inverter			
Controlled inverter				
Cut-in wind speed (m/s)	2.5			
Cut-out wind speed (m/s)	12			
Generators	Full-permanent magnetic synchronous generator			
Tower height (m)	6			
Tower diameter	60*3 steel pipe	76*3 steel pipe	76*3 steel pipe	89*3 steel pipe
Rotor power coefficient	0.40			0.42

No. 2.8

Company name	Panasonic Co., Ltd.	Application pavilions	Japan Pavilion
Company website	http://panasonic.co.jp		
Key products and services	Panasonic Group is a global electronics company engaged in the production, sale of various electrical products. The displayed Garden lamp of Wind and PV hybrid power employ the vertical axis wind-driven generator. The main features are as follows:		
	(1) It can adapt to frequent change of the wind direction and wind speed		
	(2) Have a good stability, namely at low wind speed the wind turbines can be started and in a low average wind speed area the gross generation is better than the general vertical axis wind-driven generator		
	(3) Well safety designs can bear the wind speed of 60 m/s and have special brake designs		
	(4) Special appearance different from the ordinary horizontal axis windmill can increase the visual attraction and beautify landscape		
	(5) Erection height is low, so it is convenient to installation and maintenance		
	(6) No noise. Design of vertical axis completely solves the problem of horizontal axis windmill		
	(7) It is fit for the wind regime on the roof of high buidlings		
Patents	Self-independent patent		

No. 2.9

Company name	Shanghai Ture Enclosure Electric Co., Ltd.	Application pavilions	Pavilion of Future
Company website	http://www.turebox.com		
Key products and services	Shanghai Ture Enclosure Electric Co., Ltd. is the only agent for Finland-based ENSTO Turebox in China, providing the professional solutions of seal box for customers		
	Main products: ENSTO Wind power control box, IP65 DC header box, ENSTO wind power junction box. ENSTO Cubo (box) in wind power field has following advantages:		
	(1) *High degree of protection*:		
	Reach IP67 (degree of protection), IK08 (degree of impact resistance)		
	Excellent degree of protection can cut be restricted from damage and corrosion from harsh climate. Because of the outstanding anti-impact properties, the controller can be reused and the working life can reach over 30 years without degradation or deformation		
	(2) *Leading technology of control box*:		
	ENSTO dedicated electrical control box (ENSTO Cubo) in wind power field has reached the world's advanced level on anti-flaming, anti-corrosion, water proof, dust proof, acid-proofing, alkali-proofing, shock proof, RHOS, and so on. The solution of the control box is the most complete, satisfying the need of different clients in different industries		

(continued)

(continued)

	(3) *Full range of sizes and a wide choices*:
	As one of the biggest suppliers for world's electric stuffing box, the company has over 500 sizes to choose and many kinds of materials such as cast aluminum, PC, ABS, polyester, 304 stainless steel, painting steel, and so on
	(4) *Certificates*:
	It possesses various certificates from Europe and North America such as CE, FIMKO, UL, and can provide all kinds of authoritative certificates in China
	(5) *Rich experience in customization:*
	It has rich experience on great project of construction and installation. However customer service you need, our service is the best
	Characteristics of IP65 DC header box: water-proof, dust proof, acid-proofing, alkali-proofing, shockproof, blast-proof, anti-UV, non-conducting, widely used Hostile environments such as outdoor docks, Ports, outdoor power distribution project, communication, mechanical equipment, chemical industry, petroleum industry
	The range of water proof, dust proof is IP66/67
Patent	Self-independent patent

No. 2.10

Company name	Shanghai Forevoo Windpower Technology Co., Ltd.	Application pavilions	World Expo Ferry
Company website	http://www.turebox.com		
Key products and services	Forevoo is an innovation-based high-tech company focusing on wind power field		
	In 2008, Forevoo creatively developed the dual wind rotor with overrunning clutch wind turbine, called "Double Flying Birds"		
	This product is able to capture the energy of low speed winds. The starting-up wind speed is further reduced to breeze level, while the adjustability to a broad range of wind speed is strengthened. It also has safety index with pleasant outlook which make it particularly suitable to the wind-solar hybrid power road lamp system		
	Starting-up wind speed is 1.2m/s minimum, cut-in wind speed is 1.8m/s;		
	High-efficient power generation in low wind speed;		
	Be applicable in broader scope: presenting high efficient in 80% of land area on earth		
	Safe and reliable: able to withstand Richter-13 levels typhoon		
	The "Venturi Effect Wind Turbine with Short Vanes" technology has been applied to the wind turbine series of the "Wind Ball" Model. 40 % more wind energy can be captured than traditional wind turbines with the same size of orthographic projection. It has the advantage of low wind starting up, producing energy in a wide scope, safety, and no noise		

(continued)

(continued)

Patent	Self-independent patents for "dual wind rotor with overrunning clutch wind turbine" technology and The "Venturi Effect Wind Turbine with Short Vanes" technology		
Specifications of "Wind Ball" type wind turbines			
Mode	*FDL1.4-0.3/9*	*FDL2.0-0.5/10*	*FDL2.5-1/10*
Rated power (W)	300	500	1,000
Rated voltage (V)	24	24	48
Rotor diameter (m)	1.4	2.0	2.5
Start-up wind speed (m/s)	1.5	1.6	1.8
Rated wind speed (m/s)	9	10	10
Security wind speed (m/s)	40	40	40
Yaw mode	Mechanical		
Shell material	Aluminum		
Blade material	Aluminum/Carbon fiber		
Main blade number	6	6	6

No. 2.11

Company name	Shanghai Aeolus Wind power Technology Co., Ltd.		Application pavilions	India Pavilion, World Expo Mobile Base Station
Company website	http://www.sawt.com.cn			
Key products and services	Shanghai Aeolus is an innovative wind energy technology company, entering Shanghai in 2005, engaged in the research and development on vertical axis wind turbines ranging from aerodynamics, to materials, structures, generators, and other aspects, and has made breakthrough in vertical axis wind turbine project, which remains the world-class problem			
	The Vertical axis wind-driven generator of 200 W, 300 W, 500 W, 1,000 W, 3,000 W, 10 kW has come into sales. The company has started the design and manufacture of Wind turbines with variable “angle of attack” of 50 kW, and meanwhile starts out the mechanism design of products of 500 kW and 1,000 kW			
Patents	The company has applied for 9 international patents, among which there are 3 Systematic patents for large on-grid wind-driven generator, 3 functional patents and 2 methodological patents			
Products and models	Off-Grid System: P-200, P-3000, PK-10			
	Grid-Connection System:P-1000-G, P-3000-G, P-5000-G, PK-10-G			

Parameters of wind turbines with the model of P-200, P-3000, PK-10, P-1000-G, P-3000-G, P-5000-G, PK-10-G

Model	P-200	P-3000	PK-10	P-3000-G	P-1000-G	P-5000-G	PK-10-G	Remarks
Diameter of wind wheel	80 cm	300 cm	600 cm	300 cm	180 cm	400 cm	600 cm	P-300, P-500, P-1000
Working wind speed		4–25 m/s				4–25 m/s	4–25 m/s	4–25 m/s
4–25 m/s		4–25 m/s	4–25 m/s			P-3000 is used in the India Pavilion		
Safety wind speed	40 m/s	50 m/s	55 m/s				50 m/s	
45 m/s		50 m/s	55 m/s					
Output voltage	DC 24V/ 12 V	DC 48 V	DC 110 V				DC 110–400 V	
DC 150–400 V		DC					110–400 V	DC 90–340 V
Types of brake		Automatically					Manually	Manually
Manually	Manual						Automatically	Manually
Mass of wind turbine (without tower)	48 kg	562 kg	1,905 kg				633 kg	
85 kg		985 kg	1,905 kg					
Height of wind wheel	36 cm	360 cm	620 cm				360 cm	
200 cm		460 cm	620 cm					
Rated wind speed	10 m/s	12 m/s						
Rated power	200 W	3,000 W	10 kW			3,000 W		1,000 W
5,000 W		10 kW						
Types of speed limit		Electronic		Pneumatic		Electronic		Pneumatic
Height of tower post	5.5 m							

A.3 Biomass Companies and Organizations

(See No. 3)

No.	Company name	Website	Pavilions
3	ENN Energy	http://www.enn.cn	China Pavilion, Shanghai Case Pavilion, Hydrogen refueling Station

No.3

Company name	ENN Energy	Application pavilions	China Pavilion, Shanghai Case Pavilion, Hydrogen refueling Station
Company Website	http://www.sawt.com.cn		
Key products and services	ENN Energy provides customers with personalized intelligent integral low-carbon energy solutions relying on its innovation and integration of the systematic energy efficiency technology and low-carbon energy technology with their own independent intellectual property rights. Through a diversified business portfolio with smart low-carbon energy service as the core and new business models, it meets customer's needs in energy saving and emission reduction. The main services that ENN provides include the consultancy on the integral intelligent low-carbon energy solutions, services on implementation and operational management etc. Meanwhile ENN also provides municipal, industrial and commercial clients with technical or management services in power outsourcing, centralized heat supply. The intelligent integral low-carbon energy solutions include low-carbon energy technical solution, intelligent energy efficiency service system solution, capital funding, commercial, and operational management solutions. These services help solve the problems of technology, capital, policy support, benefit sharing, and operational management system in energy saving and emission reduction and in the process of renewable energy utilization		
Patents	Independent patent		

A.4 Fuel Cell and Lithium Battery Companies and Organizations

(See No. 4, 4.1 and 4.2)

No.4

No.	Company name	Website	Pavilions
4.1	Panasonic Corporation	http://panasonic.cn	Japan Pavilion
4.2	Thunder Sky Battery Limited	http://www.thunder-sky.com	London's UBPA Case

No.4.1

Company name	Panasonic Corporation	Application pavilions	Japan Pavilion,
Company website	http://www.panasonic.cn		
Key products and services	Panasonic Corporation is a global electronic company engaged in the production and marketing of a wide range of electrical products. The ENE-FARM green eco-friendly home fuel cell can provide 60 % of the electricity for the common household with the output power of 1 kW, and reduce 37 % of the CO_2 emission. The product can guarantee a minimum 40,000 h operation, 10 years of service life, with more than 4,000 times start-up		
	Every ENE-FARM green eco-friendly home fuel cell system enjoys 45 % of the government subsidy		
Patents	Independent patent		

No. 4.2

Company name	Thunder sky Battery Limited	Application pavilions	London's UBPA Case
Company website	http://www.thunder-sky.com		
Key products and services	Thunder Sky Battery Limited specializes in the research and production of LFP material and Li-ion battery technology. With a big capacity for the two kinds of batteries materials, the company can provide the products to four kinds of electric vehicles such as pure electric cars, mini-bus, city bus and tour bus. It is also the ideal provider of LFP Li-ion batteries for solar energy, wind power, mobile communication stations, UPS power, electric yacht, electric ships, tour submarine, fuel vehicles, etc.		
	The company established the energy storage station for London case Pavilion, with 280 pieces of 700AH LFP batteries, and a total installed capacity of 620.72 kWh		
Patents	Independent patent		

A.5 Ground/Water Source Heat Pump Companies and Organizations

(See No. 5, 5.1, 5.2, 5.3, 5.4, 5.5, 5.6, 5.7 and 5.8)

No. 5

No.	Company name	Website	Pavilions
5.1	Beijing Huaqing Geothermal Development Co., Ltd.	http://www.dire.cn	Expo Axis, Expo Center, Expo Culture Center and Pavilion of Future
5.2	Polarbear Energy (Group) Company	http://www.polarbearenergy.cn	Germany Hamburg Pavilion, London Case Pavilion, Iceland Pavilion
5.3	Johnson Controls, Inc.	http://www.johnsoncontrols.com	Expo Axis, Expo Center, UBPA, InterContinental Shanghai Expo, 500kv Expo Transformer Substation, etc.
5.4	LS Air-conditioning System (Shandong) Co., Ltd.	http://www.lsasd.com	Construction site of Expo Village
5.5	Nanjing Fullshare Energy Co., Ltd.	http://www.fseet.com	Expo Axis, Expo Center, Expo Culture Center and Pavilion of Future
5.6	Shanghai Brown Environmental Technology Co., Ltd.	http://www.bllc.com.cn	United Nations Pavilion
5.7	Trane (Jiangsu) Inc.	http://www.trane-china.com	China Pavilion, Expo Culture Center
5.8	Dunham-Bush Yantai Co., Ltd.	http://www.dunham-bush.456114.com.cn	Apartments in Expo Village Zone B and Zone D

No. 5.1

Company name	Beijing Huaqing Geothermal Development Co., Ltd.	Application pavilion	Expo Axis, Expo Center, Expo Culture Center, and Pavilion of Future
Company web	http://www.dire.cn		
Key products and services	Beijing Huaqing Geothermal Development Co., Ltd. specializes in heat pump technology from introduction, R&D, application, market development, to system design and engineering construction management, and also engages in the exploration, design, construction, and production project research of cold/hot underground water resources		
	Technology: by using the huge amount of low-position energy stored in underground water, geothermal water, soil, river/lake water, and industrial wastewater, with small amount of electricity, to convert it into energy for use, energy efficiency conversion ratio can achieve 5:1, that is, consume 1 kW electricity can obtain 5 kW thermal power		
Patent	Independent patent		

No.5.2

Company name	Polarbear Energy (Group) Company	Application pavilion	Germany Hamburg Pavilion London Case Pavilion, Iceland Pavilion
Company web	http://www.polarbearenergy.cn		
Key products and services	Polar Bear Energy Inc is a research-based group specializing in water/ground source heat pump development in Canada. The main products are ground source heat pump, Integral water source heat pump and heat-recovery water source heat pump GR Series, GS Series		
	Application of the Hamburg House: Deborah Bell adopted "3U energy piles", "natural cooling" and other heat pump new technologies, launched capillary radiation water source heat pump unit combining the patented casing of "air conditioning-fresh air-dehumidifier" and board changing, and earned the affirmation of the Germany Hamburger House for its outstanding technological advantages, which are suitable for built-up city or regional projects without adequate pipe		
Patent	Independent patent		

No. 5.3

Company name	Johnson Controls, Inc.	Application pavilion	Expo Axis, Expo Center, UBPA, InterContinental Shanghai Expo, 500 kv Expo Transformer Substation, etc.
Company web	http://www.johnsoncontrols.com		
Key products and services	Johnson Controls is one of the largest global manufacturers of independent Heating Ventilating and Air conditioning (HVAC) and refrigeration equipment, ranking top 200 among the global top 500 companies. In 2005, Johnson controls acquired York International and became the world's largest supplier of product, system, and service integration. The company mainly designs, manufactures large and medium industrial refrigeration equipment		
	Main products: large and medium centrifugal and screw water chillers, industrial refrigerating machines and auxiliary machines		
Patent situation	Screw water chillers, ground source heat pump		

No. 5.4

Company name	LS air-conditioning system (Shandong) Co., Ltd.	Application pavilion	Construction site of Expo Village
Company web	http://www.lsasd.com		
Key products and services	LS Air-conditioning, detached from South Korea's LG Group in 2005, owns 17 subsidiary companies including LS-MTRON, LS Cable, LS machinery, LS Industrial Systems, LS Nikko Copper, Gaon Cable, E1, Yesco. In 2005, It changed its name to LS Group. In August 1997, LS Group invested and built factories in Qingdao, China and began to provide the central air-conditioning equipment with world's advanced technology to the Chinese market		
	LS Group produces a full range of central air conditioning products—direct-fired lithium bromide absorption chiller, direct-fired lithium bromide absorption chiller with three functions, steam lithium bromide absorption chiller, coking industry-specific type of refrigerator, hot water lithium bromide absorption chillers, flooded screw chillers, dry screw chillers, centrifugal chillers, fan coil units, modular air conditioners, air-cooled screw chillers (heat pump), air-cooled scroll chiller (heat pump), ice storage centrifugal chillers, and so on		
	Prospect: widely used in textile, pharmaceutical, tobacco, metallurgy, chemicals, hotels, hospitals, movie theaters, stadiums, office buildings, etc, with a vast market, and development prospects		
Patent	*Screw Chillers*		

No. 5.5

Company name	Nanjing Fullshare Energy Co., Ltd.	Application pavilion	Expo Axis/Expo Center, Expo Culture Center, and Pavilion of Future, Shanghai Case Pavilion
Company web	http://www.fseet.com		
Key products and services	Nanjing Fullshare Energy is a new energy service organization which engages in the ground source heat pump system technology R&D. Taking green building and energy-efficient building as its service field, based on water source heat pump, the company integrates solar cooling, building envelope, replacement of fresh air, capillary radiation, ice storage, and many other new technologies to build energy-saving, high comfort green building. In the area of energy management building, it invests in the world's largest single combination of renewable energy in scale utilization in regional energy station—Nanjing International Service Outsourcing Industrial Park zone cooling and heating supply project. Taking advantage of the natural water source from the Yangtze River, using river water source heat pump, ice storage, transmission and distribution of large temperature difference, and other advanced technologies to achieve regional cooling and heating, peak load shifting, ease the pressure on regional power. In the field of green building research, Fullshare Energy set up the first green building Engineering Research Center in Jiangsu Province		
	Application in the Expo: through eco-technology to show the characteristics of the pleasant temperature in the Expo Axis. By use of the pile foundation under this huge public channel laying 700 km long of pipelines at the bottom, making the ground-source heat pump		
Patent situation	*Independent patent*		

No. 5.6

Company name	Shanghai Brown Environmental Technology Co., Ltd.	Application pavilion	United Nations Pavilion
Company web	http://www.bllc.com.cn		
Key products and services	Brown, known as the world's air quality improvement and air energy application experts, introduced the advanced IAQ air deplacement technology and CLIMATE SYSTEM as well as the air energy heating technology to China		
	Product: Whole-house air exchange systems, whole-house climate system, Space Island, whole-house hot water system		
	Key technology: IAQ air displacement technology, CLIMATE SYSTEM building climate system technology, air energy heating technology		
Patent	IAQ air displacement technology, CLIMATE SYSTEM building climate system technology, air energy heating technology, centralized dust absorption system, IAQ central air-conditioning		

No. 5.7

Company name	Trane Inc.	Application pavilion	China Pavilion, Expo Culture Center
Company web	http://www.trane-china.com		
Key products and services	Trane was established in 1913. Since its establishment, Trane has been committed to developing more efficient, energy-saving, environmentally friendly air-conditioning products, and systems application techniques. Its products are widely used in electronics, industrial, commercial buildings, supermarket chains, financial institutions, government, high-end residential, cultural, educational, and other fields		
	Product: Roof machine, air-cooled ducted system, water source products, air-cooled chillers, water cooled chillers, air-side end, air treatment system		
Patent	*Ice-storage system, river water source heat pump system, end-equipment*		

No. 5.8

Company name	Dunham-Bush Yantai Co., Ltd.	Application pavilion	Serviced apartment in Expo Village Zone B and Zone D
Company web	http://www.dunham-bush.456114.com.cn		
Key products and services	Established in 1894 as a professional manufacturer of heating, ventilating & air- conditioning units and refrigeration equipment, Dunham-Bush is Pioneer of rotary screw compressor technology for air-conditioning systems, they enjoy a total of 48 world patents		
	Products: screw rotor units, centrifugal units, cooling tower, screw compressor, centrifugal compressor, unit machine, end-equipment		
Patent	*Air-cooled heat pump system*		

Appendix B
Companies and Organizations in Application of Building Energy Efficiency Technologies in the Expo

B.1 Energy-Efficiency Doors and Windows Companies

(See No. 1, 1.1, 1.2, 1.3, 1.4 and 1.5)

No. 1

No.	Company name	Company web	According pavilion
1.1	Shandong Jin Jing Technology Co., Ltd.	http://www.cnggg.cn	Expo Axis, Theme Pavilion, China Pavilion, Expo culture Center
1.2	CSG Holding Co., Ltd.	http://www.csgholding.com	Expo Culture Center, Belgium, UAE Pavilion, Arab Pavilion, Pavilion of Future, City Power Plant, Pavilion of Footprint, UBPA
1.3	AGC Flat Glass Co., Ltd. (Japan)	http://www.agc-flatglass.cn	Australia Pavilion, Brazil Pavilion, Spain Pavilion, Malaysia Pavilion, Japan Pavilion etc.
1.4	Xinyi Glass Holdings Limited	http://www.xinyiglass.com	China Pavilion
1.5	Zhongli Holding Group	http://www.zlholding.com	Italy Pavilion

UNIDO-ISEC and Jiangsu Modern Low-carbon Technolgy Research Institute (eds.),
The Research Report on Application of Low-carbon Technology in Expo 2010 Shanghai,
DOI 10.1007/978-3-662-44357-6

No. 1.1

Company name	Shandong Jin Jing Technology Co., Ltd.	Application pavilion	Expo Axis, Theme Pavilion, China Pavilion, Expo Culture Center
Website	http://www.cnggg.cn		
Key products and services	Jin Jing (Group) Co., Ltd. takes the development, production, processing, and marketing of glass, soda ash and extended product as its main business, and gradually extends into the fields of the solar energy new materials, new energy-saving materials. In 1998 it passed ISO9002 quality system certification. In 2007 it obtained ISO/14001 environment system certification, and 3A certification on standardization of enterprise. The company mainly produces 2–25 mm ultra-white glass, coated glass, solar energy glass, electronic glass, color glass, glass, motorcycle, Locomotive glass, float glass series, ultra-white glass, solar low iron glass, high-quality white glass, Low-E coated glass, Low-E energy-efficient insulating glass white glass, Low-E insulating glass, ultra white Low-E insulating glass,, color Low-E insulating glass, solar control coated glass		
	Ultra-white glass was provided to the Expo. Ultra-white glass is a low-iron glass, also known as low-iron glass or high-transparent glass. It is a high-quality, versatile new type of high-end glass, with light transmission rate of above 91 %		
Patent	Independent R&D center on works glass, with own patent		

No. 1.2

Company name	CSG Holding Co., Ltd.	Application pavilion	Expo Culture Center, Belgium pavilion, UAE Pavilion, Pavilion of Future, City Power Plant, Pavilion of Footprint, UBPA
Website	http://www.csgholding.com		
Key products and services	CSG Holding Co., Ltd. is the first Low-E energy-saving glass and double silver Low-E glass manufacturer. Its main products are:		
	Clear float glass, colored float glass, low emissivity coated glass (Low-E glass), heat reflection coated glass, insulating glass, laminated glass, laminated glass patterns, glazed glass, bent glass, tempered glass, curved glass, silver mirror glass, fire resistant glass and the composite glass products formed by the above glass		
	Low-E coated glass, heat-reflective coated glass, insulating glass, laminated glass were used In the Expo		
	Characteristics of Low-E glass: its energy efficiency is reflected in its shield of the sun thermal radiation (heat-insulating property) and the block of heating leakage (heat retaining property). Low-E glass is divided into two series as shading type and high-transparent type according to the different regions for use and design requirements		
	High visible light transmittance—the appearance of the permeability effect is good, and indoor natural lighting effect is perfect		
	High solar transmittance—the solar radiation through the glass is more, glass shading coefficient Sc ≥ 0.5		
	High infrared reflectivity—lower heat transfer coefficient U value, excellent insulation properties		
	The insulation characteristics that make it suitable for cold northern regions, it allows solar heat radiation into the room in the winter to increase the indoor heat energy, and send back indoor heating, home appliances and far infrared heat of the body to the room, thus to reduce the heating energy consumption effectively		
Patent	Glass works with independent R&D center, with their own patent		

No. 1.3

Company name	AGC Flat Glass Co., Ltd. (Dalian)	Application pavilion	Australian Pavilion, Brazil Pavilion, Spain Pavilion, Malaysia Pavilion, Japan Pavilion etc.
Website	http://www.agc-flatglass.cn		
Key products and services	Founded in 1992 AGC Flat Glass (Dalian) Co., Ltd. is a wholly owned subsidiary company of the Japan's AGC Group which is a Global Fortune 500 Company.		
	Utilizing the most advanced technologies that AGC Group has in float glass manufacturing, its products are mainly the special glass used in building and vehicles		
	For construction used glass: there is Solar Control Low-E Glass (Sunergy®): The online Solar Control Low-E coated glass have the functions of sun-shading and thermal insulation, it can keep a comfortable temperature of the house all the year round		
	It has the following advantages:		
	1. beautiful natural color of the appearance		
	2. with both excellent solar control performance and low-E performance		
	3. low reflectance, high transmittance		
	4. easy to process, and can make all kinds of glass deep processing		
	5. when form hollow, do not require removal the film of the edge		
	6. easy to store, bestow the storage of common float glass, do not need special treatment		
	The AGC Float Glass after deep processing has been widely used in automotive, construction, and industrial fields		
Patent situation	With their own patent		

No. 1.4

Company name	Xinyi Glass Holdings Limited	Application pavilion	China Pavilion
Website	http://www.xinyiglass.com		
Key products and services	Xinyi Glass Holdings Limited (Xinyi Glass) was founded in 1988, It is one of the main manufacturers in the global glass industry Chain. Xinyi Glass offers products and solutions covering float glass, ultra-clear photovaltaic glass, auto glass and architectural glass, etc.		
	LOW-E insulating laminated glass provided by Xinyi Glass used in China Pavilion is a new energy-saving glass, It can block most of the sun heat and UV and also has excellent sound effects and safety performance. In the aspects of energy-saving and environmental protection, it caters to the new ideas of low-carbon and global sustainable development of China Pavilion		
	The size of Low-E energy-saving glass for China Pavilion is 30 % larger than the average energy-saving building glass, its thickness has increased by nearly 50 %, almost reaching the limit of the global processing of advanced equipment. The Low-E energy-saving glass on the outer wall area of China Pavilion is about 20,000 m^2		

(continued)

(continued)

	Performance:
	1. thermal properties: low V values (thermal coefficient), filled with inert gas is more effective;
	2. the optical properties: according to actual needs, it can select different projection sunlight and reflection flexibility;
	3. noise properties: hollow glass generally reduces noise 30 dB, but when filled with inert gas, it can reduce 5 dB on the original base
	4. anti-condensation properties: the dew point is lower than the state-set −40 °C, and lower than −65 °C, to ensure that no condensation of insulating glass;
	5. leak-proofness: aluminum frame (interval frame) shaped one time, butyl rubber, and structure rubber or polysulfide rubber make the double way sealing to ensure the sealing of insulating glass and extend the life of insulating glass
Patent	With their own patent

No. 1.5

Company name	Zhongli Holding Group	Application pavilion	Italian Pavilion
Website	http://www.zlholding.com		
Key products and services	Zhongli Group was founded in 1998, It is an enterprise that integrates developing, manufacture and marketing, dedicated in green energy-efficient building materials. Currently it has a number of subsidiaries such as Zhongli Energy Efficient Glass, Zhongli PV Curtain Manufacturing, Zhongli Energy Efficient Doors and Windows, Zhongli Real Estate, Xiaofu Trading, Xiaofu Logistics, etc.		
	Main energy-efficient products: Solarban 60 double silver Low-E glass, Sungate 100 single silver Low-E glass, insulating glass, laminated glass, Sun-shading tempered glass		
	The glass products are widely used in all kinds of building glass curtains and indoor separation		
Patent	With their own patent		

B.2 Energy-Efficient Materials Companies and Organizations

(See No. 2, 2.1, 2.2, 2.3, 2.4, 2.5, 2.6, 2.7, 2.8, 2.9, 2.10, 2.11, 2.12, 2.13, 2.14 and 2.15)

No. 2

No.	Company name	Company web	Pavilion
2.1	Akzo Nobel (China) Co., Ltd	http://www.akzonobel.com/cn	10 theme Pavilions incl. Spain Pavilion, Germany pavilion, Swiss Pavilion, etc.
			Expo Axis, Expo Village, and Expo Culture Center
2.2	BASF (China) Co. Ltd.	http://www.basf.com/	Hamburg House
		http://www.greater-china.basf.com/	
2.3	Beijing New Building Materials (Group) Co., Ltd.	http://www.bnbmg.com.cn	China Pavilion, Expo Theme Pavilions, Expo Axis, Expo Culture Center
2.4	Henkel Adhesives Co., Ltd.—Shanghai Office	http://www.henkel.cn	Expo Culture Center
2.5	Nippon Paint (China) Co., Ltd.	http://www.nipponpaint.com.cn	Singapore Pavilion, Shanghai Case Pavilion
2.6	Suzuka International (Shanghai) Co., Ltd	http://www.cosuzuka.com	Expo Culture Center
2.7	Pang Beijie Paints Trading (Shanghai) Co., Ltd	http://www.nexaautocolor.com.cn	China Pavilion,Expo Theme Pavilion
2.8	PPG Industries, Inc.	http://www.ppg.com	France Pavilion
			The USA Pavilion
2.9	Shikoku Kaken (Shanghai) Co., Ltd.	http://www.skk.com.cn	Expo Culture Center
2.10	Shanghai Aux Paint Co., Ltd.	http://www.shoikos.com	China Pavilion
2.11	Shanghai Porlun Special Coating Co., Ltd.	http://www.porlun.com	China Pavilion
2.12	Shanghai Luxurious Paint Co., Ltd.	http://www.likexing.com	Expo Axis
2.13	Shanghai Yada Special Paint Co., Ltd.	http://ww.adash.com.cn	Expo Axis
2.14	Shanghai Yuan Sheng Waterproofing Engineering Co., Ltd.	http://www.ys-engineering.com	Expo Culture Center
2.15	Asia Paint (Shanghai) Co., Ltd.	http://www.asia-paint.com	China Pavilion

No. 2.1

Company name	Akzo Nobel (China) Co., Ltd.	Application pavilion	10 theme Pavilions incl. Spain Pavilion, Germany pavilion, Swiss Pavilion, etc.
			Expo Axis, Expo Village, and Expo Culture Center
Website	http://www.akzonobel.com/cn/		
Key products and services	Headquartered in Amsterdam, the Netherlands, Akzo Nobel is one of the Global Fortune 500 companies and has consistently ranked first (12 years in a row) in the global paint sales. Axe Nobel has Implemented GB18581-2001 in China while launched the "Tim gorgeous" wood coatings series to China, the oil-based paint by using the production process and technology of furniture paint		
	"Tim gorgeous" wood coating series include single-, two-component, from primer to finish, from matte to high-light, and with imported "fine color" to allow wooden furniture out of the monotony. It has obtained ISO9001 international quality system certification, ISO14001 environmental management system certification, the products meet GB18581-2001, China's latest standard for harmful substances limit in product line interior solvent-based coatings		
Patent situation	Independent patent		

No. 2.2

Company name	BASF (China) Co. Ltd.	Application pavilion	Hamburg House
Website	http://www.basf.com/		
	http://www.greater-china.basf.com/ (China)		
Key products and services	BASF is the world's leading chemical company. Its portfolio ranges from oil and gas to chemicals, plastics, performance products, agricultural products and fine chemicals. The main products of BASF Chemical Group (Guangdong) Coating Co. Ltd., include interior wall paint, exterior wall paint and wood exterior paint		
	Exterior wall paint: BSF002-DEVIS Almighty alkali-resistant primer, BSF2000-DEVIS advanced weather exterior wall paint, BSF3000-DEVIS acrylic self-cleaning exterior wall paint, BSF4000-D EVIS mildew UV protected paint walls, BSF5000-DEVIS acrylic barometer wall paint, BSF6000-DEVIS Nano flexible wall paint, BSF2001-DEVIS external alkali primer, BSF3001-DEVIS all-around external mold alkali primer		
	The new pigment technology of XfastBlack0095 and SicopalBlackK0095 not only applies to the coatings industry, but also can be used for window frames or car interior plastic parts, etc. The products of the exterior paint of BASF Chemical Group (Guangdong) Coating Co., Ltd. are suitable to the outdoor wall painting of large high-grade office, senior clubs, residential villas, and hotels		
Patent	Independent Patent		

No. 2.3

Company name	Beijing New Building Materials (Group) Co., Ltd.	Application pavilion	China Pavilion, Expo Theme Pavilion, Expo Axis, Expo Culture Center
Website	http://www.bnbmg.com.cn		
Key products and services	Beijing New Building Materials (Group) Co., Ltd. is affiliated to China National Building Material Group Co., Ltd. under the management of SASAC. In 1979 the company was invested by the state to become the largest industrial base of new building materials. Now it has developed into a conglomerates which combines investment, development of wood products, logistics, trade, and integrated housing in one, with the total assets of more than 60 billion Yuan, sales of over $30 billion		
	The product category include: building decorative materials, wood products, metallurgical materials, mineral products, machinery and equipment, chemical products, steel structured housing and general contracting services in engineering design		
Patent situation	Independent patent		

No. 2.4

Company name	Henkel Adhesives Co., Ltd.	Application pavilion	Expo Culture Center
Website	http://www.henkel.cn		
Key products and services	Founded in 1876, Henkel operates worldwide with leading brands and technologies in three business areas: Laundry and Home Care, Cosmetics/Toiletries, and Adhesive Technologies		
	Henkel leads in the global market in the civilian category, craftsman class and adhesives in industrial class, sealants and surface treatments		
	Adhesives and sealants system is widely used in automotive, packaging, aerospace, electronics, durable goods, and metal industries, and also used for maintenance, repair, and overhaul. Construction adhesives provided for architects and building contractors; Adhesives and sealants provided for the artisans and the general consumer, can be used for homes, schools, and offices		
Patent situation	Independent patent		

No. 2.5

Company name	Nippon Paint (China) Co., Ltd.	Application pavilion	Singapore Pavilion, Shanghai Case Pavilion
Website	http://www.nipponpaint.com.cn		
Key products and services	Nippon is a well-known paint manufacturer which was founded in 1881. It has a wide range of businesses, involving a variety of fields, architectural coatings, automotive coatings, general industrial coatings, coil coatings, powder coatings		
	Nippon Paint has passed the ISO9001 International Quality Management System certification and ISO14001 international environmental management system certification of Det Norske Veritas (DNV),which is one of the world's most three authoritative certification body		
Patent situation	Independent patent		

No. 2.6

Company name	Suzuka International (Shanghai) Co., Ltd.	Application pavilion	Expo Culture Center
Website	http://www.cosuzuka.com		
Key products and services	SUZUKACOAT is an international company which has been engaged in architectural coating and EIFS		
	The brand of Suzuka was established in Japan in 1945 and has made great progress in Taiwan. In 2001, Suzuka International (Shanghai) Co., Ltd. was founded in Shanghai		
	Architectural Coatings: lacquer, pottery sand aggregate, elastic coating, fluorocarbon metallic paint, exposed water concrete, interior and exterior paint, re-laminated film series		
	Insulation system: Linglu EPS exterior insulation systems, Linglu XPS exterior insulation system, Linglu insulation decorative composite panel system		
	Renovation of the old wall system: renovation of the old wall		
	Others: K1 reactive oxygen sterilization paint, 2 # 369 inorganic mineral green paint		
Patent situation	Independent patent		

No. 2.7

Company name	Pang Beijie Paints Trading (Shanghai) Co., Ltd.	Application pavilion	China Pavilion, Expo Theme Pavilion
Website	http://www.nexaautocolor.com.cn		
Key products and services	Pang Beijie Paints Trading (Shanghai) Co., Ltd was established in 1999,It is the wholly owned subsidiaries of PPG Industries in China's automotive refinish and light industrial business. Currently, the company operates two brands of PPG and NEXA AUTOCOLOR. There are a variety of products of Dazhuang, 2K, high-Ba, Haocai, Verbatim		
	Main products: Industrial anti-corrosion paint, ship and building paint, the exterior wall paint series, interior wall paint series, floor paint series, light anti-corrosion coating series		
Patent situation	Independent patent		

No. 2.8

Company name	PPG Industries, Inc.	Application pavilion	France pavilion The USA pavilion
Website	http://www.ppg.com		
Key products and services	Founded in 1883 with the headquarters in Pittsburgh, USA. PPG Industries is a global supplier of paints, coatings, optical products, specialty materials, chemicals, glass, and fiber glass		
	At present PPG has 125 factories and research institutes, located in 20 countries, is the world's leading manufacturer of coatings in aerospace, automotive and other transportation paint, construction, industrial and packaging		
	PPG's Main products involve architectural coatings, packaging coatings, automotive coatings, aerospace materials, light industrial coatings, architectural glass, glass fiber, Transitions photochromic lenses, automotive powder varnish technology, automotive noise/vibration damping coatings, high corrosion truck chassis electrophoresis paint, second-generation anti-scratch varnish nanotechnology		
Patent	Independent patent		

No. 2.9

Company name	Shikoku Kaken (Shanghai) Co., Ltd.	Application pavilion	Expo Culture Center
Website	http://www.skk.com.cn		
Key products and services	Shikoku Kaken (Shanghai) Co., Ltd. (SKK) is a wholly owned subsidiary of Japan's leading manufacturer of architectural coatings SK Kaken Co., Ltd. in China. The Product series include the general architectural coatings, coating materials, led, broad areas of activity related to other special coatings, industrial coatings, flooring and roofing materials coated materials, fire protection, thermal insulation coatings, solvents and other chemical compounds, etc.		
Patent	Independent patent		

No. 2.10

Company name	Shanghai Aux Paint Co., Ltd.	Application pavilion	China Pavilion
Website	http://www.shoikos.com		
Key products and services	Shanghai Aux Paint Co., Ltd. was founded in 1993, co-invested by Shanghai Huili Group Corporation and the Italian paint company OIKOS, specializing in the production and sale of Austrian Marcos family paint, and other coatings and construction of new services. Major equipment, production technology and processes are imported from Italy's OIKOS company		

(continued)

(continued)

	OIKOS company broke through the traditional concept of paint, and has developed a brand new water-based decorative paint system in the field of decorative coating. The company commissions a dozen varieties in seven categories the exterior decorative paints, both are water-based, including fantasy (E, ES), middle paint and primer, Austria Marcos flexible paint, Austria can be Sri Lanka metallic paint, the Austrian Marcos crack in the coating, Austria Marcos paint, Austria Marcos (1000,288,250)-type paint, Milan retaining wall paint, latex paint Duo Fenda, Duo Fenda (168,118)-based latex paint, paint the walls of Milan, Symphony (P, M, G, GS), stars (z, K), varnish (WH, WO), etc.	
Patent situation		Independent patent

No. 2.11

Company name	Shanghai Porlun Special Coating Co., Ltd.	Application pavilion	China Pavilion
Website	http://www.porlun.com		
Key products and services	Shanghai Porlun Special Coating Co., Ltd. is the a wholly owned subsidiary of Shanghai Yongli Real Estate (Group) Co., Ltd, is committed to become a high-quality domestic and international integrated supplier of corrosion resistant materials and technical service providers		
	Shanghai Porlun Special Coating is the only product distributor in China authorized by the US POR-15® company. POR-15 is a new high-performance polyurethane anti-corrosion coating produced by POR-15 company in the USA, it is a complete preservation system made up of primer, a variety of paint, high temperature paint, all kinds of metal surface treatment agent, adhesives, repair ointment. They are widely used in aerospace, defense, chemical, metallurgy, oil and gas exploration, automotive, marine, road and bridge, industrial machinery, environmental equipment, and industrial floor areas		
Patent situation	Independent patent		

No. 2.12

Company name	Shanghai Luxurious Paint Co., Ltd.	Application pavilion	Expo Axis
Website	http://www.likexing.com		
Key products and services	Shanghai Luxurious Paint is a member of Shanghai Kaile Group, devoted in producing the "Likexing" brand medium and top grade paint and fluorocarbon paints		
	Fluorocarbon carbon paint: JF "Likexing" Building walls fluorocarbon paint, JF aluminum imitation effect fluorocarbon, JF Steel fluorocarbon, JF heavy duty fluorocarbon		
	Exterior Latex paint: JW Senior exterior latex paint, JW flexible exterior latex paint, JW acrylic exterior latex paint, JW exterior latex paint		
	Interior Latex: JN senior Interior Latex, JN Acrylic Interior Latex		
	New products: JD water-based floor paint, Wood, JD water-based metallic paint, JF water-based fluorocarbon, JZ natural lacquer		
Patent situation	Independent patent		

No. 2.13

Company name	Shanghai Yada Special Paint Co., Ltd.	Application pavilion	Expo Axis
Website	http://www.adash.com.cn		
Key products and services	Shanghai Yada Special Paint Co., Ltd. was founded in 1993, in the development, manufacturing, sale and service of the paints. Currently, Yada has an annual capacity of 20,000 tons coatings, 50,000 tons of special external insulation dry mortar, 5 million m^2 of special anti-fire insulation phenolic board, 5 million m^2 of light granite decorative integrated product boards, and 2 million m^2 wall decoration board. The products cover the insulation decoration integrated system, exterior insulation systems, exterior paint, floor paint, wall coating plates, luminous paint and interior paint, and other health paints		
	Yada Special Fire phenolic board can be used in thin plaster external wall insulation system, Fire Barrier, curtain wall system, activities of the board color steel sandwich panels, fire doors, and fire insulation for special occasions. Compared with rock wool and glass wool, it has the advantage of simple, more health, environmentally friendly. Yada special fire phenolic board does not melt or drip under high temperature, or softening, It has low smoke, non-proliferation flame, resistance of flame penetration, with thermal conductivity coefficient of 0.035 W/(m · K)		
Patent	Independent patent		

No. 2.14

Company name	Shanghai YuanSheng Waterproofing Engineering Co., Ltd.	Application pavilion	Expo Culture Center
Website	http://www.ys-engineering.com		
Key products and services	Shanghai YuanSheng is a waterproof, thermal insulation energy efficient company integrating the research, design, construction, technical advice and sale as a whole. It is		
	Waterproof: The company, through its independent R&D, produces and sells the "Hufang-brand" types of waterproofing materials: 991 polyurethane waterproof coating, ShiTean high elastic waterproof coating, one-component water solified polyurethane waterproof coating, waterproof elastic acrylic coating, 951 color flexible waterproof coating, JS waterproof coating, cementitious capillary crystalline waterproofing coating, polyurethane sealing agent, the water does not leak, water treasure, self-adhesive membrane, self-adhesive polyester fetal membrane, APP, SBS waterproofing membrane, poly ethylene polymer composite waterproofing membrane (polypropylene, polyester) and so on		
	Insulation: The company independently develops and markets powder series products: polystyrene board adhesives, polystyrene board Plastering mortar, powder particles of polystyrene insulation paste, cracking mortar, mortar interface, the interface of concrete waterproofing agent, flexible adhesives, tile adhesive, stone adhesives, flexible caulking agent		
Patent	Independent patent for water-proof paints		

No. 2.15

Company name	Asia Paint (Shanghai) Co., Ltd.	Application pavilion	China Pavilion
Website	http://www.asia-paint.com		
Key products and services	Asia Paint (Shanghai) Co., Ltd. is the world's leading provider of comprehensive coating solutions. It is based on the paint industry since its initial production in 1998, and involved in a number of areas and related areas with the paint industry, while providing comprehensive solutions for high-quality coating system-level products		
	Asia thermal insulation technology: provide integration of insulation and finished decorative plate insulation system for the full coating solutions		
	Asia paint: provide high-end users with the outstanding comprehensive painting and decoration services		
Patent situation	Independent patent		

B.2.1 Other New Type of Building Materials

In the Shanghai Expo, a number of pavilions displayed some kinds of new energy-saving materials, some of which are made of wheat straw or used CD, and other materials are wastepaper. It is true that their prospect remains to be seen, but it indicates a direction for human beings as how to reduce the energy consumption. These materials and their applications in the Expo are listed below:

(See No. 2.16, 2.17, 2.18, 2.19, 2.20, 2.21, 2.22, 2.23, 2.24, 2.25, 2.26, 2.27, 2.28, 2.29, 2.30, 2.31, 2.32, 2.33, 2.34 and 2.35)

No. 2.16

Applied technology	Building materials decoration		
Company name	A.O. Smith Corporation	Application pavilion	China Pavilion, Expo Axis, Expo Culture Pavilion
Website	http://www.aosmith.com.cn		
Key products and services	A.O. Smith Corporation is a world's renowned US-based company. In 1998, A.O. Smith US headquarters invested solely to establish the A.O. Smith (China) Water Heaters Co., Ltd. in Nanjing, China		
	The Company has an excellent modern management system on R&D, production, sales, and service integration, the product system covers three major areas of industrial, commercial, and household use		
	Household type: EWH-E5 EWH-D5 HPW CEWH-PEZ5 JSQ-G1 JSQ-E/EX JSQ-C2/C2X SWHN EMGP-C EMGO-C		
	Commercial type: Commercial gas water heater/furnace: BTR/BTRO HW GW/GB DW/DB Commercial electric water heaters: DRE/DSE DEN NW Commercial Water softener: CSIR		
	Applications: Hotels, factories, enterprises, government offices, beauty/fitness/leisure, living district heating, catering industry		
Patent	Independent patent		

No. 2.17

Company name	UPM Beijing office	Application pavilion	Finnish Pavilion (scale-type wall board)
			China Pavilion (Section III)
Website	http://www.upm-kymmene.com.cn/		
Key products and services	Founded in Finland in 1900s, UPM is one of the leading forestry industrial groups. Its products are made of renewable raw materials and are recyclable. UPM comprises of three Business Groups: Energy and pulp, Paper, and Engineered materials		
	UPM ProFi wood plastic composite materials:		
	It is a unique material that combines the advantages of wood fibers and plastic, made by recycled material, which is proven to be tough and humidity resistant. UPM ProFi mainly uses the wastes generated in the process of self-adhesive label production as the raw material. its main		

(continued)

(continued)

	component is paper and plastic. The composite material can be disposed by incineration or recycled back into the production process. The advantages are:
	1. Good strength properties; 2. water, corrosion resistance, longer service life; 3. Excellent scalability; 4. With ultraviolet light stability, good coloring; 5. The biggest advantage is turning waste into something valuable, and 100 % recovery of reproduction; 6. it can be degraded and will not cause the "white pollution", which is regarded as the real green products; 7. A variety of sources of raw materials; 8. the product can be made into any shape and size as needed
Patent	Independent patent

No. 2.18

Applied technology	Building materials decoration		
Company name	Hangzhou Nabel Group Co. Ltd.	Application pavilion	Expo Culture Center
Website	http://www.nabel.cc		
Key products and services	Hangzhou Nabel is a foreign-funded company, set up in Hangzhou, China in 1992. As a frontrunner of the ceramic tiles industry in China, it has the world's advanced production line and the cutting edge equipment imported from Italy, e.g., Furnace, ROTOCOLOR, ROLLFEED and Italy Kerajet digit printing, which represents the world's leading technology. It mainly products are fully vitrified tiles, matt tiles, retro European-style brick, Class ceramic composite plates, ceramic glazed wall and floor tiles, ceramic glazed wall and floor tiles, and other decorative accessories, such as the "Nobel" and "Cezanne • Impression" series of products		
Patent	Applied patent: 338 items, authorized patent 282 items		
	Applied invention patent: 4 items, 1 is approved		

No. 2.19

Applied technology	Straw board pressed by wheat straw		
Company name	Netherlands Leenderson	Application pavilion	Vanke Pavilion
Website	http://www.leenderson.com		
Key products and services	Netherlands Leenderson (China) Alfredsson board Holdings (China) Co., Ltd. based in The Hague, Netherlands, has signed with Canada Edmonton, Alberta Innovates Technology Futures of Edmonton an exclusive license agreement on global production of wheat straw OSB board (OSSB) and medium density fiberboard (MDF). Its China headquarters and marketing center is located in Beijing, the first directional OSSB board production line is in Yangling, Shaanxi (west of Xi'an)		
	Netherlands Alfredsson board industry uses environmentally friendly OSB OSSB board, combined with cold-formed light steel structure system (LGS) and Western construction methods, to create a way of income for Chinese farmers, through avoiding the burning of straw and reduction of the use of brick and cement, significantly reduces carbon dioxide emissions. As the sole supplier of straw board pressed by wheat straw to Vanke Pavilion of the		

(continued)

(continued)

	Expo, the Netherlands Alfredsson board industry is the world first to produce zero-emission formaldehyde OSB board
	Characteristics of Oriented Structural Straw Board (OSSB)
	Load bearing, distinct from traditional boards in the physical properties
	Environmental friendly materials with formaldehyde emission close to zero
	Excellent nail holding strength
	Aesthetic from inside to outside
	Good fire resistance
	The straw board is pressed by natural wheat straw
	It provides a new effective way to reuse of crop straw, and reduce consumption of forest resources
Patent	OSSB board is developed by the Alberta Innovates Technology Futures of Edmonton, Canada and acquired the patent

No. 2.20

Company name	Ningbo, Weishan Duobao Building Material Co. Ltd.	Application pavilion	China Pavilion
Website	http://www.duobau.com		
Key products and services	Ningbo, Weishan is a professional supplier of glass fiber mesh products. The company started to enter the fiberglass industry in 1989, and officially registered as Cixi Weishan chemical Building Materilas Co., Ltd. in 1997. In 1999, Ningbo Weishan Du Bao Building Material Co., Ltd. was established		
	Products: Fiberglass Mesh, Exterior insulation and Enhanced mesh, self-adhesive fiberglass mesh, Mosaic mesh lining, Angle bar mesh, Roof tarps, wall enhanced mesh, Alkali fiberglass mesh, GRC enhanced mesh, granite marble back paste mesh		
Patent situation	Fiberglass mesh		

No. 2.21

Applied technology	Building insulation		
Company name	Owens Corning (China) Investment Co., Ltd.	Application pavilion	China Pavilion, Expo Axis, Expo Culture Center
Website	http://www.owenscorningchina.com		
Key products and services	Owens Corning based in Toledo, Ohio, USA, It is leader in world's building materials and glass fiber composites fields		
	Products: glass wool insulation, extruded foam board, vinyl decorative wall cladding, artificial stone exterior materials and Silentex system, other composite materials for solutions to automobile industry		
Patent	Glass wool and XPS extruded plastic board materials		

No. 2.22

Company name	Shanghai Chen Jie Engineering Plastics Co. Ltd.	Application pavilion	China Pavilion, Expo Axis
Website	http://www.cjsl.cnjg.com.cn		
Key products and services	The company is a processing enterprise that is specialized in producing general level and flame-retardant grade EPS products. It can produce various sizes of EPS products with the density ranging 15–40 Kg/m^3. Large size board of high density is used in road infrastructure		
	Environment protection and energy-saving technology: Geothermal heat system, External wall insulation system, etc.		
Patent situation	EPS board		

No. 2.23

Company name	Shanghai Chengnuo Brothers Industrial Co. Ltd.	Application pavilion	Expo Axis
Website	http://www.ji-di.com		
Key products and services	Chengnuo Brothers is a professional enterprise specialized in producing Jidi brand copper fittings for advanced bathrooms. Products: basin drain, Bathtub drain, a variety of inlet pipe, angle valve, flush value, small tap, floor drain, piping, and other eight series, a total of 160 kinds of products		
Patent situation	Independent patent		

No. 2.24

Company name	Shanghai Caoyang Construction Adhesives Factory	Application pavilion	China Pavilion, Expo Culture Center, Theme Pavilions
Website	http://www.newcaoyang.com		
Key products and services	Shanghai Caoyang Construction Adhesives Factory was established in 1988, It is a professional manufacturer of the premixed polymer dry mortar, starting early in the field of dry powder in China. The technology in production and product quality reached international advanced level		
	"Shen Tai" brand JCTA products: series of light block, Interface treatment series, building Waterproof Series, Series of adhesives and sealants, Tile and mosaic processing series, self-leveling, flooring disposal Series, Building interior and exterior putty series, dried powder pre-mixed mortars, Internal and external wall insulation series		
Patent situation	Inorganic insulation mortar, putty, adhesives, epoxy floor commercial mortar		

No. 2.25

Company name	Shanghai Huayuan Composite New Materials Co. Ltd.	Application pavilion	Expo Theme Pavilion
Website	http://www.huayuanfu.com/		
Key products and services	The company is a joint venture, specialized in producing Huayuan aluminum Plastic composite plate (APCP), copper Plastic composite plate, Titanium plate, Stainless steel plastic composite board, Galvanized steel-plastic composite board, Bimetal Strip, and matching silicone rubber		
Patent situation	APCP, copper-plastic composite plate		

No. 2.26

Company name	Shanghai Lafarge Gypsum Co., Ltd.	Application pavilion	Expo Axis, Expo Culture Center
Website	http://www.lafarge-gypsum.cn		
Key products and services	Lafarge has the world's leading building materials research and development facility. Lafarge Gypsum Building Materials entered the Chinese market in 1996. The company is headquartered in Shanghai, and has 4 plants in Shanghai, Chongqing and Chengdu, with an annual output of 100 million m^2 gypsum board and ancillary products. The company mainly provides gypsum board system products for the domestic interior architectural design and decoration market, including gypsum board walls, ceiling, wall, and ceiling series of decorative veneer. Lafarge adopts the world's most advanced production equipment and technology, the system made of gypsum board and related parts belongs to the new type light, high-strength, environmental protection and energy-saving decorative materials, with excellent thermal insulation, sound insulation, fire protection, moisture-proof performance, and various decorative effects. Lafarge also offers a series of solutions to gypsum board system design and installation		
Patent	Sensium® Cement products, with dust-proof technology		

No. 2.27

Company name	Shanghai Louxin New Building Materials Co. Ltd.	Application pavilion	China Pavilion, Expo Culture Center, Expo Axis
Website			
Key products and services	Aerated concrete hollow brick		
Patent	Independent patent		

No. 2.28

Company name	Shanghai Pudong Yi Xiang, Insulation Materials Co. Ltd.	Application pavilion	Expo Culture Center, Theme Pavilion
Website	http://www.shyxbw.com		
Key products and services	Shanghai Pudong Yi Xiang Insulation Material Limited company is specialized in producing all kinds of EPS insulation board and XPS Extrusion board. The product has the perfect obturator honeycomb structure. Its water absorption rate, thermal conductivity coefficient and permeability coefficient and other key parameters are less than other insulation materials, therefore it has the advantages of high strength, light weight, corrosion resistance, aging resistance, low prices, widely used in various occasions. The company has a complete set of automated control of the EPS polystyrene insulation board production line, XPS Extrusion board production line		
Patent	EPS board, XPS board		

No. 2.29

Company name	Shanghai Lei Meng Energy-saving Building Materials Co. Ltd.	Application pavilion	China Pavilion
Website	http://www.lei-meng.com		
Key products and services	Lei Meng Energy-saving Building Materials Co. Ltd. focuses on R&D, production, sales, and service of external wall thermal insulation product, applicable to various areas of external building walls. In order to better meet the insulation requirements of building exterior wall in different climate regions, the company developed a variety of high quality exterior wall insulation products, and formed different series as:		
	thermal insulation mortar for exterior wall thermal insulation system, polystyrene board exterior insulation system, heat insulation decoration board system, the insulation inside the external wall, roof waterproof, and insulation system		
Patent	External wall insulation mortar		

No. 2.30

Company name	Shanghai CIMIC Tile Co, Ltd.	Application pavilion	China Pavilion, Expo Culture Center
Website	http://www.cimic.com/		
Key products and services	The Kilns are SACMI, SITI, NIRO, and STUDIO One		
	Presses are SACMI and SITI		
	Glazing line is SYSTEM ROTO-COLOR		
	Green body porcelain machine is CMF		
	Loading Machines are B&T and CMF		
	CIMIC is a Sino-British Joint venture. It is one of the largest tile manufacturers in China specialized in producing and marketing advanced polished porcelain tiles and ceramic tiles. It has a variety of product portfolio which is in more than 20 series, and 10 specifications, almost a hundred in number for flagship products. Glazed ceramic tiles have 6 sizes, including wall tiles, floor tiles, tiles, belts, corner thousands of		

(continued)

(continued)

	products. Most CIMIC products have the technology superior to the national standard. CIMIC's self-developed products, the external wall hanging, hanging paste products and technologies as well as anti-static flooring enjoy wide praise in and outside the industry
	Wall hanging paste technology has gained the Enterprise Standard Certification from Ministry of Construction for building products and components, fittings Product Standardization Technical Committee. Its insulation and energy-saving wall hanging magnetic board is the initiative in China
Patent situation	Independent patent

No. 2.31

Company name	Shanghai Sonata New Wall Material Co., Ltd.	Application pavilion	China Pavilion, Expo Axis
Website	http://www.sonnt.com		
Key products and services	Shanghai Sonata introduced the world's most advanced air turning six surface cutting production technology, specializing in the production of autoclaved lightweight concrete block and plate sonata		
	The Sonata lightweight aerated block takes quartz sand as the basis, with cement and lime as cemedin, gypsum as hardening agent, aluminum powder as foaming agent, at high temperature (180–200 °C) and high pressure (10–13 atm) conservation for 8–12 h, this porous light wall is formed		
	The features of the Sonata lightweight aerated concrete block:		
	Light weight:It weighs about red-brick 1/3 or concrete 1/4. It can effectively reduce the weight of the building while reducing investment in 27 % of the base, then the total cost can be reduced by 5 %		
	Good energy saving: the insulation performance of 40 mm thick wall can be comparable to the wall with red brick 240 mm thick. It is an excellent auto-wall insulation material		
	Excellent fire performance: the fire resistance of the brick with 200 mm thick is up to 8 h		
	Excellent sound insulation: Sonata aerated block wall can be separated more than 52 dB sound		
	Excellent impermeability: When using ordinary external painting, the impermeability of the walls is 85 % higher than the brick wall's		
	Excellent use: As a relative small thickness of the product, it can increase the actual area of the house		
	Green environmental: Sonata Light sand aerated concrete block is inorganic. It does not burn and produce toxic gases, what's more, it isn't radioactive at all		
Patent situation	Sonata lightweight aerated block		

No. 2.32

Company name	Shanghai SEALOP Building Materials Co., Ltd.	Application pavilion	Expo Culture Center
Website	http://www.sealop.com		
Key products and services	Shanghai Sealop is a chemical construction material production company with International advanced technology, Since it was founded in 2003, the company has been devoted to the research and development, production and sale of the building insulation, waterproofing and adhesion and other chemical building materials, with a strong R&D capability and perfect service. They have "SEALOP", "SEALQUIC" and other registered trademarks in domestic and global market. They have also passed the ISO9001:2008 international quality management system certification		
	Main products: Exterior Insulation and Finishing System (EIFS), Underground, Bathroom waterproofing system, Roofing insulation system, Interior and exterior wall putty, Land grant system of color finishes, Advanced ceramic tile adhesive, Color sealant, Floor leveling system		
Patent situation	Advanced ceramic tile adhesive		

No. 2.33

Company name	Shanghai Yuanxia Decoration Materials Co., Ltd.	Application pavilion	China Pavilion, Expo Axis
Website	http://www.yuanxia.net		
Key products and services	Shanghai Yuanxia has, cooperated, since 1997, with Japan XiZhiye Chemical Corporation, OWINGS (the sole strategic partner of BEYER in China), Malaysia HUME and a number of famous domestic and foreign enterprises and successfully developed a new generation of multi-functional decorative plates Portland—Cable Kit board, which has filled in the gap in China. Cable clean plate production process and application technology are in the advanced level in our country		
	Water board, Polyester plates, Fluorocarbon board, Antimicrobial board, Granite plate, Inorganic board, underground space system, clean system, external wall system		
Patent	Cable kit board		

No. 2.34

Company name	Tohoku University	Application pavilion	National Information Development Pavilion
Website	http://www.tohoku.ac.jp		
Key products and services	Following Tokyo University (1877) and Kyoto University (1897), Tohoku University was founded in 1907 as the third Imperial University. Today it has become a large-scale comprehensive national university		
	Its developed product Low-temperature ceramics: low-cost, low waste incineration ash to achieve safe disposal; as a high-strength solid its strength is 3–7 times of concrete; weight is about 2/3 of concrete; better than concrete in waterproofing; used to make colored products. Effective use of incineration ash to extend the life of landfill; incineration incomes from ash treatment and sale of products to increase profits rate. Clay, construction sludge, stone charcoal, broken glass, waste insulator and other wastes can be used as raw materials		
	Wide applications: building materials, civil engineering materials, marine materials		
Patent	Independent patent		

No. 2.35

Company name	Xinyi Ceramics (China) Co. Ltd. Shanghai Branch	Application pavilion	Expo Axis
Website	http://www.sinyih.com.cn		
Key products and services	In 1972, the founder of Mr. Lin Tiancun established Xinyi Ceramics Industry Co. Ltd. In Taiwan. In 1997, Xinyi ceramics (China) Co. Ltd. was set up in Kunshan, Jiangsu. It aims to build a park-like, automated, zero emission production base of green tiles. By introducing the world's top Italian equipment, it produces world class quality products		
	Main products: Nano Antibacterial, self-cleaning, anti-skid ceramic tiles etc.		
Patent	Independent patent		

B.3 Building Afforestation Enterprises and Organizations

(See No. 3, 3.1, 3.2, 3.3 and 3.4)

No. 3

No.	Company name	Website	Pavilions
3.1	Hainer Roofing system	http://www.hainer.cn	Canada Pavilion, Ireland Pavilion
3.2	Firestone Building Products United States	http://www.firestonebpe.com	India Pavilion
3.3	Shanghai Yaxuan green plant nutrition Soil factory	http://www.yingyangtu.com	China Pavilion, Expo Theme Pavilion
3.4	SOPREMA, Inc.	http://www.soprema.cn	Italia Pavilion

No. 3.1

Company name	Hainer Roofing System	Application pavilion	Canada Pavilion
Website	http://www.hainer.cn		
Key products and services	Heiner Roofing Systems Company is the earliest and the most professional in China in producing, designing, installation, maintenance roofing, roof insulation, building green overall system, —i.e. the roof Systems (Roofer). It has the most advanced PVC/TPO/SBS waterproofing membrane production line, and it is an excellent contractor for the Shanghai Expo in numerous pavilions roof system design and building		
	Green Roof System		
	Hainer green roof system includes a light green roof (soil layer: 5–25 cm) and heavy green roof (soil: ≥25 cm)		
	Especially for concrete, light steel, wood and other smaller capacity provided by Hainer light green roof: excellent waterproof, ultra-light weight, quick construction, simple maintenance. Simple structure: vapor barrier, insulation, Heiner GF block membrane-type plant roots, soil, plants (Note: The heavy green roof should be based on need for additional drain and filter cloth.) Fixation can be mechanically fastened, adhesive pressure fixed or empty shops re-fixed		
Patent	Independent patent		

No. 3.2

Company name	Firestone Building Products United States	Application pavilion	India Pavilion
Website	www.firestonebpe.com		
Key products and services	Firestone Building Products is part of the Bridgestone Corporation, the world's largest tire and rubber company, mainly in the production and marketing of a range of commercial and residential roofing and lining systems. It is Headquartered in Indianapolis, Indiana, USA, its products include ethylene propylene diene monomer rubber (EPDM) membranes, thermoplastic membranes, modified bitumen, poly-isocyanurate insulation, and related supporting material		
Patent	Independent patent		

No. 3.3

Company name	Shanghai Yaxuan Green Plant Nutrition Soil Factory	Application pavilion	China Pavilion, Expo Theme Pavilion
Website	http://www.yingyangtu.com		
Key products and services	Shanghai Yaxuan Green Plant Nutrition Soil Factory, professionally produces the roof afforestation light soil (green roof substrate), green nourishment soil, saline soil improvement of the media etc. Factory direct produce nutritive soil, perlite, vermiculite, Longkang ash, peat Northeast, Northeast peat, imported peat, coco peat, pine needles soil, cooked chicken and other gardening materials		
	Green Private Peat Northeast (peat soil): Saline-alkali soil improvement medium, city, highway, lawn planting, tree transplanting, garden nurseries, roof garden, potted flowers, and improved activation of the soil, improve soil organic matter content. Green roof light nourishment soil: Dedicated green roof light nourishment soil (light soil). Green roof light soil, roof garden light soil, rooftop green light soil		
	Green Project nutritive soil: city, highway, lawn planting, tree transplanting, improving and activating the soil, improve soil organic matter content		
	Fine fiber peat (turf): fruit and vegetable seedings, soil improvement, tree transplanting, landscaping, greening projects		
	Advanced lightweight green roof nutritive soil		
Patent situation	Independent patent		

No. 3.4

Company name	SOPREMA, Inc.	Application pavilion	Italia Pavilion
Website	http://www.soprema.cn		
Key products and services	SOPREMA (Thorpe Rama) Group was founded in France in 1908, it is one of the international group in the R&D, production, design, sale, construction, and service of the modified bitumen materials (SBS, APP), FLAG (Farag) polymer waterproof material (TPO, PVC) and related auxiliary materials. After years of development and innovation, Soprema Group provides a range of high-quality systematic solution for different users. Soprema roof garden system was developed in 1982, which has a complete systematic concept and the corresponding solution for different conditions. As the system saves resources, being environment friendly, it is promoted and applied in Europe first, and has been successfully applied in more than 1,000 projects around the world		
Patent	Independent patent		

Appendix C
Companies and Organizations in Application of Other Energy-Saving Technologies in the Expo

C.1 List of LED Companies and Organizations

(See No. 1, 1.1, 1.2, 1.3, 1.4, 1.5, 1.6, 1.7 and 1.8)

No. 1

No.	Company name	Company website	Pavilion
1.1	PAK Electrical Appliance Co., Ltd	http://www.pak.com.cn	One Axis and Four Pavilions
1.2	Royal Philips Electronics	http://www.lighting.philips.com.cn	Shanghai Case Pavilion
1.3	Osram (China) Lighting Co., Ltd.	http://www.osram.com.cn	China Pavilion, Expo Center and the "We are the World' Pavilion, streets, pavements, and tourist attractions
1.4	Advanced Photonics. Co., Ltd	http://www.shapc.com.cn	Israel pavilion, Indonesia Pavilion, Expo Center, Saudi Pavilion
1.5	Shenzhen Jansun Lighting Co., Ltd	http://www.dengjuled.com	Expo Garden and some pavilions
1.6	General Electric Company	http://www.ge.com	U.S. Pavilion
1.7	ZheJiang JingRi Lighting Technology Co., Ltd	http://www.jingrilight.com	Zhejiang Pavilion, Expo Plaza
1.8	Zhejiang Mingchuang Optoelectronic Technology Co., Ltd	http://www.mc-led.com	Pavilion of Future

UNIDO-ISEC and Jiangsu Modern Low-carbon Technolgy Research Institute (eds.),
The Research Report on Application of Low-carbon Technology in Expo 2010 Shanghai,
DOI 10.1007/978-3-662-44357-6

No. 1.1

Company name	PAK Electrical Appliance Co., Ltd	Application pavilion	One axis and four pavilions
Website	http://www.pak.com.cn		
Key products and services	PAK Electrical Appliance Co., Ltd., which was founded in 1991, has been devoting itself to development and production of green energy saving lighting products of high quality. PAK has developed indoor, outdoor, decorative lighting, and electrotechnical products, including a variety of lighting fixture, lamp, electrotechnical product, and electric fittings totally over 2,000 categories		
	Characteristics of PAK LED:		
	(1) the use of imported high-quality high-power LED light source, light source has a long life, low lumen depreciation, maintenance, etc		
	(2) matching precision optical lens, to achieve accurate and uniform light distribution effects, the built environment to meet the different		
	(3) compact and flexible form factor structure of the patent, combined with professional design of the cooling fins, and can effectively transfer heat from the light source to the case, extend life LED light source		
	(4) red, green, blue, white, four different LED light colors, combined with the variable replaced by RGB color changes more		
	(5) controlled by DMX, lighting effects editing to meet customer requirements		
	(6) With built-in control circuit and power supply means to reduce the voltage control signal and the light color cast caused by the gap dist		
Patent	Independent patent		

No. 1.2

Company name	Royal Philips Electronics	Application pavilion	Shanghai Case Pavilion
Website	http://www.lighting.philips.com.cn		
Key products and services	Royal Philips Electronics is a diversified company which serves professional and consumer markets through three overlapping sectors: Healthcare, Lighting and Consumer Lifestyle Healthcare, Lighting and Consumer Lifestyle		
	Related to light, professional lighting, LED lighting, lighting electronics, lighting control		
	Applied in offices, shops, hotels, medical health organizations, industry, and outdoor lighting		
Patent situation	Independent patent		

No. 1.3

Company name	Osram (China) Lighting Co., Ltd.	Application pavilion	China Pavilion, Expo Center and the "We are the World" Pavilion, streets, pavements and tourist attractions
Website	http://www.osram.com.cn		
Key products and services	OSRAM, headquartered in Munich, Germany, is a wholly owned subsidiary of Siemens. At this Expo, OSRAM energy efficient lighting solutions created the best lighting effects in the various buildings within the park, streets, trails and tourist attractions. OSRAM's finest works include China Pavilion, Expo Center and the "World" pavilion. In this Expo, over 150,000 LED lights of Osram were used in various pavilions and streets in the Expo park		
Patent situation	LED built-in chip technology		

No. 1.4

Company name	Advanced Photonics. Co., Ltd.	Application pavilion	Israel pavilion, Indonesia Pavilion, Expo Center, Saudi Pavilion
Website	http://www.shapc.com.cn		
Key products and services	Advanced Photonics. Co., Ltd. is jointly invested by the United States Prince Glory and Shanghai Jinqiao (Group) Company, specializing in the semiconductor LED chips, LED components, LED light production and development		
	The company has four major products:		
	① LED chip: annual output of 3,000 KK, with yield and quality in the leading position		
	② LED components: production of various high-end LED devices 60 KK, sells well at home and abroad		
	③ LED indoor and outdoor single, two-color screen and full-color display: The company specializes in producing indoor and outdoor LED full color, all kinds of LED graphic display		
	④ LED lighting products		
	The application of the Expo: Providing Israel Museum, Museum Indonesia, Saudi Arabia Pavilion with outdoor full color LED display, the Expo Center with floodlighting project, large transparent display, bar display for the Theme Pavilion		
Patent	Digital LED color light source, a color light band with light-emitting diodes as a light emitting device, a high-power light emitting diode (LED) device, a light band jacket		

No. 1.5

Company name	Shenzhen Jansun Lighting Co., Ltd.	Application pavilion	Expo Garden and some pavilions
Website	http://www.dengjuled.com		
Key products and services	Shenzhen Jansun Lighting Co., Ltd. is a manufacturers of LED and LED-related products in China. It has been engaged in the research, manufacture and marketing of LED based on High-power and quality (semiconductor) Lighting		
	Jansun mainly produces high-power LED lighting, LED Landscape Lighting, LED Outdoor Lighting, LED Office lighting, LED accessories, etc. including high-power LED floodlights, LED spotlights, LED line lights, LED buried lights, high power LED spotlight, LED wall lamp, LED down light, LED underwater lights, LED lamps, high power LED lights Cup and other series of high-power LED outdoor lighting		
	The company's products have been applied in buildings, road, bridges, shopping malls, hotels, parks, museums, public facilities, and other areas for decorative lighting		
Patent	Independent patent		

No. 1.6

Company Name	General Electric Company	Application pavilion	U.S. Pavilion
Website	http://www.ge.com		
Key products and services	GE is a diversified company covering technology, media, service, and finance. GE has a large variety of products, from energy, petrol and natural gas, water treatment, aviation, transportation, appliance, lighting and medical, finance, and news media		
	GE's LED products:		
	High-power white light LED, LED logo lighting, refrigerator lighting, jewelry lighting, wall-washing lighting series as well as regional lighting system		
	Innovative high brightness R63 LED products		
	Applications: Interior design and general lighting, architectural lighting, LED advertising light font and logo lighting, indoor and outdoor directional lighting, security access indication, custom fixture and built-in lamps, LED message signs		
Patent situation	Independent patent		

No. 1.7

Company name	ZheJiang JingRi Lighting Technology Co., Ltd.	Application pavilion	Zhejiang Pavilion, Expo Plaza
Website	http://www.jingrilight.com		
Key products and services	Jingri is a high-tech company devoted to technology innovation of the energy-saving products and research on LED high-tech products projects		
	JR LED series products: LED lights, LED floodlights, LED tunnel lights, LED underground lamps, LED downlight, LED solar street lamps		
	Applications in the Expo: LED floodlights in Zhejiang Pavilion, LED street light in Expo plaza		
Patent situation	Independent patent		

No. 1.8

Company name	Zhejiang Mingchuang Optoelectronic technology Co., Ltd.	Application pavilion	Pavilion of Future
Website	http://www.mc-led.com		
Key products and services	Zhejiang Mingchuang Optoelectronic Technology Co., Ltd is the state owned High-tech enterprise Specializing in LED lighting product development, production, sales, and service. It has made breakthroughs in the core technologies of power semiconductor lighting which has achieved mature applications of 10–150 W single LED light source. The products can be widely used in roads, tunnels, factories, mines, landscapes, and commercial lightings		
Patent situation	Digital LED color light source, a color band with light-emitting diodes as a light emitting device, a high-power LED device, a light band jacket		

C.2 Non-electric Central Air-Conditioning Enterprises and Organizations

(See No. 2.1)

No. 2

No.	Company name	Company website	Pavilion
2.1	Broad Air-conditioning Equipment Co., Ltd.	http://www.broad.com/	Broad Pavilion

No. 2.1

Company name	Broad Air-conditioning Equipment Co., Ltd.	Application pavilion	Broad Pavilion
Website	http://www.broad.com/		
Key products and services	Broad is a large private enterprise. BROAD Air Conditioning Equipment Co. Ltd. supplies non-electric central air conditioning powered by natural gas and waste heat, provides packaged water distribution system as well		
	BROAD Air Quality Technology Co. Ltd. supplies central air conditioning terminal products with electrostatic cleaners, provides heat recovery fresh air units, air purifiers, and phone air quality detectors. BROAD Energy Service Co. Ltd. provides energy management contract service, district cooling-heating-power system (CHP) investment, building and operation. BROAD Sustainable Building Technology Co. Ltd. provides factory-made sustainable buildings with Level 9 earthquake resistant, with 6 times material less, 5 times energy efficient, and 20 times air purification		
	Expo Broad Air Conditioning Co., Ltd. built 22 energy centers in the Expo area, providing non-electric central air conditioning for more than 200 pavilions		
Patent	Independent patent		

C.3 Energy Saving Air-Conditioning Enterprises and Organizations

(See No. 3, 3.1, 3.2, 3.3, 3.4, 3.5, 3.6, 3.7, 3.8, 3.9, 3.10, 3.11, 3.12, 3.13, 3.14 and 3.15)

No. 3

No.	Company name	Company website	Pavilions
3.1	Daikin Industries, Ltd.	http://www.daikin-china.com.cn	Japanese Industry Pavilion, Taiwan Pavilion
3.2	Guangdong Jirong Air-conditioning Co., Ltd.	http://www.jirong.com	Pavilion of Footprint
3.3	Guangdong Euroklimat Air-Conditioning & Refrigeration Co., Ltd.	http://www.euroklimat.com.cn	Indonesia Pavilion, Egypt Pavilion, Kazakh Pavilion, etc.
3.4	Guangdong Shenling Air-conditioning Equipment Co., Ltd.	http://www.shenling.com	Expo site
3.5	Guangdong Chigo Air Conditionning Co., Ltd.	http://www.china-chigo.com	Expo Puxi Admin. Center
3.6	HAYE Group	http://www.haye.cn	Pudong Logistics Expo Center
3.7	Multistack	http://www.superlink-china.com	Romania Pavilion, Turkey Pavilion, Ningbo Case Pavilion
3.8	Climaveneta ChatUnion Refrigeration Equipment (Shanghai) Co., Ltd.	http://www.climaveneta.com.cn	Italy Pavilion
3.9	Midea Group	http://www.midea.com.cn	Expo Area A09, A10 hall, Expo Village Apartment Hotel
3.10	OAK Central Air Conditioning	http://www.oakchina.com	Canada Pavilion
3.11	Mitsubishi Heavy-Haier	http://www.mhaq.cn	Expo site Zone D Senior Apartment Hotel
3.12	Mitsubishi Heavy Industries Air-conditioners	http://www.mhi-ac.com	Korea Pavilion
3.13	TCL Corporation	http://www.tcl.com	Expo temporary venues and facilities Zone C1, Expo Hotel Shanghai East King
3.14	Tecka	http://www.tecka.com	Saudi Pavilion
3.15	Dunan Mechanical and Electrical Technology Co., Ltd.	http://www.sh-dunan.com	Germany Pavilion, Switzerland Pavilion

No. 3.1

Company name	Daikin Industries, Ltd.	Application pavilion	Japanese Industry Pavilion, Taiwan Pavilion
Website	http://www.daikin-china.com.cn		
Key products and services	Daikin Industries, Ltd. is a Japanese company with its business covering a number of areas including air-conditioning, refrigeration, fluorine chemistry, electronics and hydraulic machinery, etc.		
	Daikin's air-conditioning products: home use central air-conditioners, home split type air-conditioners, heating and hot water systems, air cleaners, VRV central air-conditioners, air-conditioners for equipment use, water chillers, etc.		
	Application at the Shanghai Expo: VRV III system used in Japanese industry Pavilion and Taiwan Pavilion		
Patent situation	Independent patent		

No. 3.2

Company name	Guangdong Jirong Air-conditioning Co., Ltd.	Application pavilion	Pavilion of Footprint
Website	http://www.jirong.com		
Key products and services	Guangdong Jirong Air-conditioning Co., Ltd. (hereinafter "Jirong"), founded as early as 1985, is located in Jieyang City, Guangdong Province, China. After more than 25 years' development since its establishment, Jirong has strived to become a large company specializing in central and industrial air conditioners, which integrates R&D, manufacture, sales and services		
	Jerong air-conditioned concrete products: cell-type air conditioner, roof air conditioners, water chillers, air cold (hot) water units, air handlers (including fan coil units), air conditioners for computer rooms, multi-dehumidifiers, clean air conditioners, special air conditioning (low temperature, high temperature, grain cooler), a total of nearly 2,000 varieties and specifications		
	Jirong's central air conditioners are widely used in hotels, shopping malls, and office buildings. Its industrial air conditioners are applied in airports, power plants and metallurgical factories, as well as for military and pharmaceutical purposes. Jirong has a large number of clients across the country and is exporting its products to many countries		
Patent situation	Independent patent		

No. 3.3

Company name	Guangdong Euroklimat Air-Conditioning & Refrigeration Co., Ltd.	Application pavilion	Indonesia Pavilion, Egypt Pavilion, Kazakh Pavilion, etc.
Website	http://www.euroklimat.com.cn		
Key products and services	Euroklimat Air Conditioning Co., Ltd. is a joint venture invested by Italy's EK group and China Aerospace Science and Industry Group, EK Group is Europe's leading refrigeration and air-conditioning equipment manufacturers. EK was the first to apply HFC 134a refrigerant in the water chillers without damaging the ozone layer and has maintained the advantages in the environmentally friendly refrigerants applications. The products include water-cooled chillers, air-cooled chiller (heat pump), water (ground) source heat pump, hot water units, multi-split central air conditioning units, unit-type air conditioners, terminal of central air conditioning and precision air conditioning, etc.		
Patent situation	Independent patent		

No. 3.4

Company name	Guangdong Shenling Air-conditioning Equipment Co., Ltd.	Application pavilion	Expo site
Website	http://www.shenling.com		
Key products and services	Established in 1988, Shenling is engaged in providing large-and-medium-sized air-conditioning equipment and HVAC integrated solutions for commercial, industrial and high-end civil and residential use and the integral solutions		
	Shenling products: chillers, terminal series, clean air conditioners, dehumidifiers, constant temperature and humidity air-conditioners, roof air-conditioners, ceiling air-conditioners, unit air-conditioners, special air-conditioning unit		
Patent situation	Independent patent		

No. 3.5

Company name	Guangdong Chigo Air Conditioning Co., Ltd.	Application pavilion	Expo Puxi Administration Center
Website	http://www.china-chigo.com		
Key products and services	Guangdong Chigo Air Conditioning Co., Ltd. is a large enterprise mainly engages in designing, R&D manufacturing and sale of air-conditioners and central air-conditioners		
	For residential air conditioners there are four categories: wall type, Floor standing type, window type and portable type, in more than 1,000 specifications. Central air-conditioners include CMV-[V] DC inverter multi series, air-cooled heat pump chillers, water-cooled scroll screw-type chillers, ceiling cassette type, ESP duct type, air-cooled closet type, water-cooled closet type, unitary and packaged type, American type duct, terminals, etc. Besides, it has other electrical products such as cooker, multi-function electric pressure cooker, soybean milk machine, refrigerators, washing machines, freezers, wine cooler, air-powered central water heater, etc.		
Patent situation	Independent patent		

No. 3.6

Company name	HAYE Group	Application pavilion	Expo Pudong Logistics Center
Website	http://www.haye.cn		
Key products and services	Haye Group is specialized in development, production, sale of air-conditioners and commercial hot water machines and offering the engineering services. By designing the advance integral air-conditioning system, and making use of renewable energy as a primary energy of the water/ground/air source heat pump, insulation materials and special materials to reduce energy consumption and emissions of waste gas, it has broke the dependence on traditional energy sources, providing people with overall energy-saving air-conditioning set integrating refrigeration, heating, floor heating, hot water, and air purification system		
	Air-conditioning products include central air-conditioning, commercial air- conditioners, water (ground) source heat pump units, special air-conditioners, hot water machines		
Patent situation	Independent patent		

No. 3.7

Company name	Multistack	Application pavilion	Romania Pavilion, Turkey Pavilion, Ningbo Case Pavilion
Website	http://www.superlink-china.com		
Key products and services	Multistack is an international refrigeration & air-conditioning supplier		
	The Products include special constant temperature and humidity air-conditioning units, commercial air-conditioning equipment, water source heat pump, heat pump water heaters, air handling equipment, brazed plate heat exchanger		
Patent situation	Independent patent		

No. 3.8

Company name	Climaveneta ChatUnion Refrigeration Equipment (Shanghai) Co., Ltd.	Application pavilion	Italy Pavilion
Website	http://www.climaveneta.com.cn		
Key products and services	Climaveneta ChatUnion Refrigeration Equipment (Shanghai) Co., Ltd. (CCU China), as a number of Climaveneta SPA, is a manufacture of chillers and heat pumps for air-conditioning and industry process		
	The main products cover more than 20 series in 2,000 kinds of models, including water-cooled chillers, air-cooled units, water source heat pumps, fan coil, cabinet air-conditioning unit, modular air-conditioning units		
	Air-conditioner Host Performance: Air-cooled screw heat pump by switching the four-way valve to change the direction of the refrigerant cycle, according to the user needs to provide cold water or hot water. The unit is the air-source heat pump for reverse cycle, eliminating the need for cooling towers, cooling water pumps and cooling water systems, saving cooling system investment and operating costs, can be directly installed on the rooftop or outdoor space without a special room		
	Used fro a variety of commercial and industrial purposes, meeting the needs of refrigeration and heating throughout the year		
Patent situation	Independent patent		

No. 3.9

Company name	Midea Group	Application pavilion	Expo Area A09, A10 hall, Expo Village
Website	http://www.midea.com.cn		
Key products and services	Midea Group is a large comprehensive enterprise focusing on home appliance, extending to logistics and other areas		
	Midea refrigeration products are the "air energy" central air-conditioners and home central air-conditioners in a variety of models		
	MDV multi-connected types air-conditioners, L series centrifuge were used at the Shanghai Expo		
Patent	Independent patent		

No. 3.10

Company name	OAK Central Air Conditioning	Application pavilion	Canada Pavilion
Website	http://www.oakchina.com		
Key products and services	OAK Air Conditioning Technology Co., Ltd. is a Sino-German joint venture As a low-carbon and energy-efficient enterprise, OAK has its own research Institute in developing the products, in production of large water chiller, air handling equipment, cleaning air-conditioning, home central air-conditioning, unitary air-conditioning unit, etc.		
Patent situation	Independent patent		

No. 3.11

Company name	Mitsubishi Heavy-Haier	Application pavilion	Expo site Zone D Senior Apartment Hotel
Website	http://www.mhaq.cn		
Key products and services	Mitsubishi Heavy Industries, Haier (Qingdao) Air Conditioner Co., Ltd. was jointly established by Mitsubishi Heavy Industries, Ltd. and Haier Group		
	It produces home central air-conditioning, commercial central air-conditioning, chillers, unitary series total heat-exchangers. The product is intelligent control, efficient and safe, with energy-saving ratio meeting the national standard		
Patent	Independent patent		

No. 3.12

Company name	Mitsubishi Heavy Industries Air-conditioners	Application pavilion	Korea Pavilion
Website	http://www.mhi-ac.com		
Key products and services	Mitsubishi Heavy Industries Air Conditioning Systems (Shanghai) Co., Ltd. belongs to Mitsubishi Heavy Industries Group. It focuses on the production of household air-conditioners, residential central air-conditioning, heat pump dual-loop heating system, inverter multi-connected air-conditioning, air-conditioning for shops		
Patent	Independent patent		

No. 3.13

Company name	TCL Corporation	Application pavilion	Expo temporary venues and facilities Zone C1, Expo Hotel Shanghai East King
Website	http://www.tcl.com		
Key products and services	TCL Corporation is a global business group in consumer electronics		
	TCL's air-conditioning products include wall-mounted air-conditioners, vertical air-conditioners, portable air-conditioners and dehumidifiers, etc.		
Patent	Independent patent		

No. 3.14

Company name	Tecka	Application pavilion	Saudi Pavilion
Website	http://www.tecka.com		
Key products and services	Tecka is a business group in central air-conditioning system, which combines development, manufacture, sales, and installation services in one		
	Main products: Water-cooled screw type water chiller, Air-cooled screw type water chiller (heat pump), Air-cooled scroll type water chiller (heat pump), Air handling unit, Fan coil unit, air handling unit, etc.		
Patent	Independent patent		

No. 3.15

Company name	Dunan Mechanical and Electrical Technology Co., Ltd	Application pavilion	Germany Pavilion, Switzerland Pavilion
Website	http://www.sh-dunan.com		
Key products and services	Dunan Mechanical and Electrical Technology Co., Ltd. is the wholly subsidiary of Zhejiang Dunan Environment Equipment Co. Ltd., specializing in producing electric refrigeration type central air conditioner and the terminal equipment. Currently, Duan has a variety of commercial air conditioning products in ten series and hundreds of specification, including water-cooled screw, air screw, air-cooled scroll, water source heat pumps, rooftop unit, air-cooled closet type air-conditioning and terminal equipment. Intelligent control, network remote control and energy-saving technology have been fully applied in the product		
	In the nuclear area, it designed and developed nuclear grade chillers, air handling units of nuclear grade, nuclear grade cooler, nuclear grade valve and a series of supporting equipment, making it the first nuclear grade chiller manufacturers in China supplying to the Nuclear power plants		
	Rooftop air-conditioning unit is a central air-conditioning products combining cold/hot source and air treatment into one of (also known as direct evaporative central air-conditioning). The product is usually air-cooled, through combination of various segments of cooling systems, control systems and air handling functions, to complete heat dehumidification, humidification, muffler, air purification, fresh air conveying		

(continued)

(continued)

	Dunan type water/ground source heat pump takes the underground water as the main source of energy, the electricity as the auxiliary energy, through advanced water (ground) source heat pump central air-conditioning system, explores and utilizes the inexhaustible underground energy maintained at a lower temperature all year round, to turn it into energy available to meet the demands of summer cooling and winter heating, It is a new energy-efficient, healthy non-polluting heat pump type air-conditioning system using renewable energy resources	
Patent situation		Independent patent

C.4 Ice Cooling Storage Companies and Organizations

No. 4

No.	Company name	Website	Pavilion
4.1	Carrier Corporation	http://www.carrier.com.cn	Expo Axis, Expo Center, and Theme Pavilion
4.2	Ingersoll Rand	http://www.irco.com.cn	Expo Axis, China Pavilion

(See No. 4.1 and 4.2)

No. 4.1

Company name	Carrier Corporation	Application pavilion	Expo Axis, Expo Center and Theme Pavilion
Website	http://www.carrier.com.cn		
Key products and services	Carrier has been the world biggest suppliers of air conditioning, heating and refrigeration systems and provides energy management and sustainable global leader in building services as well. Carrier Corporation is headquartered in Connecticut, USA Farmington City, its production and sale cover more than 170 countries including China		
	① commercial air-conditioning: water-cooled units, air-cooled unit, lithium bromide units, air terminal		
	② Domestic/light commercial air-conditioning: direct expansion air-conditioning units, air-cooled heat pump units, unit machine, water source heat pump units		
	③ building control; water source heat pump system, ice storage technology, water storage technology, high-efficient frequency conversion centrifugal, efficient air-cooled screw heat pump, etc., efficient, safe and economic refrigerant, retail food refrigeration systems		
Patent situation	Owns thousands of conditioning patent technology		

No. 4.2

Company name	Ingersoll Rand	Application pavilion	Expo Axis, China Pavilion
Website	http://www.irco.com.cn		
Key products and services	Ingersoll Rand is a global diversified enterprise. Ingersoll Rand's business covers industrial technology, temperature control system technology, security technology and civil systems technology		
	Ingersoll Rand Industrial Technology aims to enhance customers' energy efficiency, productivity and operational capabilities. Products include a full set of air compression systems, fluid handling systems for materials and tools and pumps, as well as hoist winch for material handling. temperature control technology that Ingersoll Rand has provides customers worldwide with efficient heating, ventilation, air conditioning and refrigeration solutions. Its brands include Hussmann, Thermo King and Trane. Ingersoll's Security Technology products include mechanical locks, electronic locks, hotel locks, mechanical and electrical integration; door closers, exit devices, steel doors and door frames and other architectural hardware products, and electronic and biometric access control systems, attendance and personnel scheduling systems, video surveillance systems, security integration systems, command centers integrated system		
	Its ice storage technology was used in China Pavilion, where there is a huge ice-storage tank under the pavilion. At night when electricity is cheaper it uses electricity to generate ice, and during the day cold water produced by ice is used to cool down the pavilion		
Patent	Independent patent		

Appendix D
Companies and Organizations in Application of Water Treatment Technologies in the Expo

(See No. 1, 1.1, 1.2, 1.3, 1.4, 1.5, 1.6, 1.7, 1.8, 1.9, 1.10, 1.11, 1.12, 1.13, 1.14, 1.15, 1.16, 1.17 and 1.18)

No. 1

No.	Company name	Website	Pavilions
1.1	Condias Gmbh	http://www.condias.de	Germany Pavilion
1.2			Chengdu's UBPA Case Pavilion
1.3	Degrémont Technologies Ltd.	http://www.degremont-technologies.com	France Pavilion
1.4	Runtech Water Treatment Company (of Russian Nuclear Technology)	http://www.runtech.ru/node/1078	Russian Pavilion
1.5	Norit Nederland BV	http://www.norit.com	Netherlands Pavilion
1.6	Solinst Canada Ltd.	http://www.solinst.com	Canada Pavilion
1.7	Kardan Water International Group	http://www.kwigwater.com	Israel Pavilion
1.8	Litree Enterprise	http://www.litree.com	Inside Expo Park
1.9	Malthe Winje DWS AS	http://www.mwdws.com	Norway Pavilion
1.10	Nitto Denko Corporation	http://www.nitto.co.jp	Osaka Case Pavilion
1.11	Shanghai Dongshuo Environmental Technology Limited Company	http://www.denovogroup.com.cn	Japan Pavilion
1.12	Shanghai Wansen Water Treatment	http://www.shwashine.com	Mainly used in London Case Pavilion
1.13	Suzhou Chuanglian Purification Equipment Co., Ltd	http://www.szclsb.com.cn	Expo Culture Center, Expo Center
1.14	Shenzhen DongFangQiSheng industrial Co., Ltd.	http://www.szdfqs.com	Shanghai Corporate Joint Pavilion, Expo Site
1.15	Dow Chemical (China) Investment Co., Ltd.	http://www.dow.com	Pavilion of Future
1.16	General Electric Company (GE)	http://www.ge.com	American Pavilion
1.17			

(continued)

UNIDO-ISEC and Jiangsu Modern Low-carbon Technolgy Research Institute (eds.),
The Research Report on Application of Low-carbon Technology in Expo 2010 Shanghai,
DOI 10.1007/978-3-662-44357-6

(continued)

No.	Company name	Website	Pavilions
	Zhejiang Suichang Xufu Bamboo Charcoal Co., Ltd.	http://www.lzztb.com	Joint Pavilion of International Organizations
1.18	Yiyuan (Shanghai) Environmental Group Co., Ltd	http://www.yytoilet.com	UN Pavilion, Life & Sunshine Pavilion, SAIC-GM Pavilion

No. 1.1

Company name	Condias Gmbh	Application pavilion	Germany Pavilion
Website	http://www.condias.de		
Key products and services	CONDIAS GmbH was founded in 2001 as a spin-off of the Fraunhofer Institute, With its founding, CONDIAS GmbH took up the production of conductive diamond electrodes		
	This diamond Electrode is used to remove harmful organic substances in the water without adding any chemical substances. This method can effectively avoid the unacceptable matter like insecticide and medicaments in drinking water		
Patent situation	Independent patent		

No. 1.2

Company name	Application pavilion	Chengdu's UBPA Case Pavilion
Website		
Key products and services	The "workflow" of "Living Water Park" in Chengdu Case Pavilion has the following steps:	
	Step 1: part of the wastewater in the Expo Park runs into the collection tank by pipeline and is pumped to the anaerobic sedimentation tank for treatment	
	Step 2: the treated water is absorbed by the physical water sculpture, inflated aerobic in a natural drop and then goes into the constructed wetland	
	Step 3: the constructed wetland was paved with a large number of pebbles, to which the attaching biological bacteria that absorbs organic matter in wastewater and purifies the water gradually	
Patent situation	Independent patent	

No. 1.3

Company name	Degrémont Technologies Ltd.	According pavilion	France Pavilion
Website	http://www.degremont-technologies.com		
Key products and services	Degrémont Technologies Ltd. is a professional water treatment company subordinated to Suez (Environment) Group, whose main business is to provide drinking water production and a variety of services for water distribution, sewage collection and treatment, waste disposal and recycling, the complete facilities and environmental protection equipment		
	Degrémont Technologies and its subsidiaries share the same brand name. These subsidiaries include:		
	Anderson Water Systems Canada: specializing in industrial process water treatment system, with the core technology as demineralization of packed layer, ion exchange, reverse osmosis, degasification, ultrafiltration and nanofiltration, etc.		
	Aquasource of France: professional ultrafiltration membrane core technology and standardized clear/disinfection technology		
	Infilco of North America: providing a full set of solutions in four main business areas of drinking water, industrial water, wastewater and sludge treatment. Core technologies include efficient clarification, filtration, biological treatment, pretreatment and high-temperature fluidized bed technology		
	Innoplana company Degremont technology: a professional sludge treatment technology and equipment company that provides sludge thickening, dewatering, drying and burning and other professional solutions		
	Ozon Virginia Company: World leader in the production of ozone and ultraviolet disinfection technology, with its business network throughout the world		
	British Triogen: a professional UV and ozone disinfection equipment and technology company whose equipment is widely used in industrial, commercial, and residential applications		
Patent situation	Patent belongs to the company		

No. 1.4

Company name	Runtech Water Treatment Company (Russian Nuclear Technology)	Application pavilion	Russian Pavilion
Website	http://www.runtech.ru/node/1078		
Key products and services	Russian Nuclear Technology Group is engaged in a series of water treatment equipment production with Runtech as the trademark, which use a lot of advanced technologies such as ultrafiltration, reflux penetration, deionization, MBR, SBR, and AF, etc.		
Patent situation	Patent belongs to the company		

No. 1.5

Company name	Norit Nederland BV	Application pavilion	Netherlands Pavilion
Website	http://www.norit.com		
Key products and services	Norit Group develops and applies the most advanced purification technology and equipment		
	The Norit Air Lift membrane bioreactor combines biodegradation process and membrane separation in the same reactor. The combination has great advantage over conventional activated sludge system, such as improving MLSS greatly, reducing the reactor volume and residual sludge highly, and saving the further treatment for residual sludge		
	Air life membrane bioreactor puts the membrane outside of the reactor, which makes the running maintenance and extension easy, and energy consumption comparable or lower than immersed MBR		
	The Netherlands Pavilion of the Expo showcased a world's smallest purification system in its HAPPY STREET, which is 1 m wide and 2 m high. With the activated carbon technology, it can produce the highest level of water. It got water with jars from the Huangpu River, through the purification offered the visitors with purified drinking water		
Patent	Patent belongs to the company		

No. 1.6

Company name	Solinst Canada Ltd.	Application pavilion	Canada Pavilion
Website	http://www.solinst.com		
Key products and services	Solinst Canada Ltd. has been in the research and manufacture of the hydrologic monitoring of groundwater, sampling technology products, and provide in-situ groundwater contamination repair solutions since 1980		
	The company focuses on two product categories:		
	level measuring device: type 101,122, 102, 103, 107 TLC;		
	groundwater Sampler, type 410, 407, 408,408M, etc.		
Patent	Independent patent		

No. 1.7

Company name	Kardan Water International Group	Application pavilion	Israel Pavilion
Website	http://www.kwigwater.com		
Key products and services	Kardan Water (KW) is the operational platform of TAHAL Group in Chinese market. Kardan Water develops, operates, and invests in water-related projects in Mainland China, focusing on water treatment and wastewater treatment, as well as water reclamation and desalination		
Patent situation	Patent belongs to Kardan Water International Group		

No. 1.8

Company name	Litree Enterprise	Application pavilion	Inside Expo Park
Website	http://www.litree.com		
Key products and services	Litree Enterprise is a professional hi-tech enterprise group specialized in Science & Technology research such as water processing, membrane separation technology research and products development, water purification equipment research and development, manufacturing, selling		
	Litree has been focused on developing various technology combination cored with ultrafiltration, successfully avoided secondary pollution in the water pipe Unlike traditional water purifier, "Litree family water purification system" ensures not only clean drinking, but also durable and clean filter cartridge. Litree products are also wildly applied in catering industry, factories and mines, schools, government units, office buildings, etc. to fulfill commercial water-use and public drinking		
Patent	Independent patent		

No. 1.9

Company name	Malthe Winje AS	Application pavilion	Norway Pavilion
Website	http://www.malthe-winje.no/		
Key products and services	Malthe Winje AS was founded in 1922, with its scope of activities covering:		
	Electro technical products Low Voltage		
	Electro technical products High Voltage		
	Automation and PLC systems		
	Smart House Solution		
	Automation/Remote Control projects within:		
	Water and Sewage Treatment		
	Power stations and distribution (High Voltage)		
	HVAC (building automation and marine automation		
	Traffic Control systems/equipment		
	Real estate development		
	Norway Pavilion showcased the solar-powered water purification equipment, which turns the contaminated rainwater into pure safe drinking water. The system is able to remove all contaminants in the water, including bacteria, viruses, parasites, etc. The whole equipment contains pre-treatment system, reverse osmosis systems, chemical cleaning systems, sterilization systems, and water supply system		
Patent	Belong to Malthe Winje AS patent		

No. 1.10

Company name	Nitto Denko Group	Application pavilion	Osaka Case Pavilion
Website	http://www.nitto.co.jp		
Key products and services	Nitto Denko rely on its core technology of adhesive technology and coating technology, attaches a variety of additional features to the membrane and sheet materials. In its factories around the world they produce and sell a wide range of electric-related and high-polymer products such as LCD optical films, auto parts, desalination membrane, transdermal drug delivery patches. The main products include double-sided adhesive tape, sealing materials, surface protection materials, construction with protective material, masking tape, bar code systems, clean room with a product, bend circuit boards, optical materials, semiconductors, electronic components, tape, fluorine resin, fluorine resin porous membrane TEMISH®, porous thin film (NITOSEP/SUN MAP) and so on		
	At the Expo it showed how to use the reverse osmosis membrane to translate the water that could not be used anymore into drinking water. Reverse osmosis membrane is a membrane with a lot of fine holes (diameter of only 1 nm), with pressure as the driving force, it can filter impurities from sea water or wastewater, thus pure water is obtained		
Patent	Independent patent		

No. 1.11

Company name	Shanghai Dongshuo Environmental Technology Limited Company	According pavilion	Japan Pavilion
Website	http://www.denovogroup.com.cn		
Key products and services	DongShuo is a high-tech enterprise with international capital investment and a provider of high-tech systems in industrial water recycling solutions as well as the general contractor		
	DongShuo is committed to research and development of the latest international membrane separation and environmental protection technology and their applications, providing customers with complete system solutions and general contracting services		
	In August 2009, it won the bidding for the pure water treatment projects in Japan Pavilion of the Shanghai World Expo, which demonstrated biological activated carbon water treatment, microbial immobilized supporter water treatment, RO membrane treatment		
Patent	Integration continuous membrane filtration device, Detach-YMQ (Patent No. ZL032309872), Detech-WOY (Patent No. ZL99213252.5)		

No. 1.12

Company name	Shanghai Wansen Water Treatment	According pavilion	London's UBPA case
Website	http://www.shwashine.com		
Key products and services	Shanghai Swanson Water Treatment Co., Ltd. was established in 1995, specialized in water treatment technology research and development as well as production of various types of water treatment equipment and water treatment chemicals, operating various types of chemicals, hazardous chemical substances. Shanghai Wansen Water Treatment Co., Ltd. is the head office of Wansen		
	Swanson's main business scope:		
	1. undertakes various water treatment projects: boiler water softening equipment, industrial water softening equipment, food industry, water softening equipment, automatic water softener, water softening plant, industrial water recycling project		
	2. production of purified water treatment equipment: by-filters of industrial circulating water, quartz sand filter, activated carbon filters, manganese sand filter, bag filter, precision filter, home central water purifier, molded filters, microporous filters and related accessories, and other supplies		
	3. provides water treatment technologies services: central air-conditioning water treatment, industrial circulating water, boiler water treatment, drinking water treatment, wastewater operation and maintenance, water treatment technologies such as ion exchange water treatment services, and ancillary equipment		
	4. production of recycled water reuse equipment: car washing water cycle purification, bathing water recycling equipment		
	5. production of disinfection dosing equipment: automatic dosing devices, automatic discharge device used in industrial circulating water treatment dosing, water treatment dosing for central air-conditioning, swimming pool water, fountains, etc.		
	6. undertakes ecological water treatment project: the establishment of river ecology and restoration and to establish residential landscape water ecology		
	Wansen water treatment technology is applied is the London Zero-carbon Pavilion, the roof rainwater was collected into one processor for processing to be for different uses to its quality. The first rainwater will be collected for toilet flush, the secondary water was used as domestic water and the final treated water after sterile was used as drinking water		
Patent	Independent patent		

No. 1.13

Company name	Suzhou Chuanglian Purification Equipment Co., Ltd	Application pavilion	Expo culture Center, Expo Center
Website	http://www.szclsb.com.cn		
Key products and services	Suzhou Chuanglian Purification Equipment Co., Ltd. is a Hi-tech enterprise, specialized in industrial water equipment, civil pure water equipment, water treatment reuse facilities, sewage treatment equipment and engineering, air purification products and engineering, design, It is a professional water treatment equipment and air purification products company combining manufacturing, sale, installation, technical service in one. The main products are deionized water equipment, cosmetics, fine chemicals water treatment equipment, electronics industry high-quality water equipment, medical equipment, water purification, high-quality direct drinking water projects, pure water equipment for surface treatment, circulating cooling water systems, river water purification equipment, water reuse systems, laboratory water purifier, desalination systems		
Patent	Independent patent		

No. 1.14

Company name	Shenzhen DongFangQiSheng Industrial Co., Ltd.	According pavilion	Shanghai Corporate Joint Pavilion, Expo Site
Website	http://www.szdfqs.com		
Key products and services	Shenzhen DongFangQiSheng Industrial Co., Ltd. pursues low-carbon economy ideas, focusing on comprehensive water treatment and water landscape art, combining development, design, manufacture, and project construction		
	Scope of activities:		
	Hydraulic air floatation circulating Fine filter, integration biological membrane reactors, General agent of membrane equipment in south China region, sewage treatment, reclaimed water recycling, river/lake/swimming pool water treatment; environment art of water landscape, music fountains, design and construction of the cooling system projects		
Patent situation	Independent patent		

No. 1.15

Company name	The Dow Chemical Company	Application pavilion	Pavilion of Future
Website	http://www.dow.com		
Key products and services	The Dow's diversified industry-leading portfolio delivers a broad range of technology-based products and solutions to customers. Its leading business in special chemical and new materials, agriculture and plastics has been applied in the fast growing market such as electronic products, water treatment, energy, coatings, and agriculture		
	Dow technology has become integral to chemical water treatment solutions, FILMTEC&S482, seawater desalination OS and reclamation processes in critical industries throughout the world		
Patent	Patent belongs to DOW		

No. 1.16

Company name	General Electric Company (GE)	Application pavilion	USA Pavilion
Website	http://www.ge.com		
Key products and services	GE's water treatment products include: GE water purifiers, GE Merlin; GE Homespring; Homesrping central water purifier; Merlin reverse osmosis machine; GE filter; the US GE water purifier; GE Modular, EDI super pure water module		
	MELIN—a product produced by US GE, using the cutting-edge membrane filtration technology, directly remove the minor impurities, suspended solids, colloids, organic matter, heavy metals, bacteria, viruses and other harmful substances from water. In the first grade, pre-activated carbon filter is used; grades 2 and 3 are the reverse osmosis (RO) membrane; in grade 4 is activated carbon filter. Continuous flow—which is widely used for direct drinking and cooking water for home, offices and schools drinking water, as well as catering industry, field camp operations, small clinics, communities, laboratories, photo processing, humidifiers, etc. It has high flow rate and low cost, available for more use points		
Patent	Independent patent		

No. 1.17

Company name	Zhejiang Suichang Xufu Bamboo Charcoal Co., Ltd.	Application pavilion	United Pavilion of International Organizations
Website	http://www.lzztb.com		
Key products and services	Zhejiang Suichang Xufu Bamboo Charcoal Co., Ltd, founded in 2000, is a leading enterprise in the research and development, manufacture, and marketing of bamboo charcoal products. It provides a convenient and eco-friendly bamboo charcoal water purifier, consists of the filtration container and a filter. The filter contains diatomite earth-ceramic filtration layer and bamboo charcoal layer installed inside the hollow groove. It is environmentally friendly, with its granular high adsorptive bamboo charcoal, effectively remove the pollutants, suspended matter and odor in the water		
Patent	Bamboo charcoal anion deodorization charcoal mud, bamboo charcoal fiber		

No. 1.18

Company name	Yiyuan (Shanghai) Environmental Group Co., Ltd,	Application pavilion	UN Pavilion, Life & Sunshine Pavilion, SAIC-GM Pavilion
Website	http://www.yytoilet.com		
Key products and services	Yiyuan (Shanghai) Environmental Group Co., Ltd, a scientific enterprise engaged in the R&D and commercialization of water-conservation and emission-reduction products		
	The "YYTOILET" toilets, based on its independent R&D, was the designated products for the United Nations Pavilion, Theme Pavilion (Life & Sunshine Pavilion) and SAIC-GM Pavilion of Expo 2010, conferred the titles of "Energy-reducing Products in Shanghai", "Recommended Products for the Experiencing Center of Scientific Expo" and "Products Promoting Energy Reduction & Environmental Protection for Shanghai Finale of 2010 MISS EARTH Green Environmental Protection"		
Patent	Water conservation patent with numerous independent intellectual property rights		

Appendix E
Companies and Organizations in Application of New Energy Vehicles in the Expo

No. 1

No.	Company name	Website	Pavilions
1.1	Anhui Ankai Automobile Co., Ltd.	http://www.ankai.com	Expo Site
1.2	BYD Company Limited	http://www.bydauto.com.cn	Expo Site
		http://www.byd.com	
1.3	Peugeot Auto	http://www.peugeot.com	France Pavilion
1.4	Chongqing Changan	http://www.changan.com.cn	Expo Site
1.5	Delft University of City Technology	http://www.tudelft.nl	Netherlands Pavilion
1.6	Venturi	http://www.venturi.fr	Monaco Pavilion
1.7	Toyota	http://www.toyota.com	Japan Pavilion
1.8	Jiangsu Xinri E-Vehicle Co., Ltd	http://www.xinri.com	Expo Site
1.9	Jiangsu Yadea Technical Development Co., Ltd.	http://www.yadea.com.cn	Expo Site
1.10	Lifan Group	http://www.lifan.com	Expo Site
1.11	Mercedes-Benz	http://www3.mercedes-benz.com	Swiss Pavilion
1.12	NAC Special Purpose Vehicle Company	http://www.yjzyc.com	Expo Site
1.13	Norway ELBIL NORGE	http://www.puremobility.com	Portugal Pavilion
1.14	Shanghai Volkswagen	http://www.csvw.com	Expo Site
1.15	SAIC Group	http://www.saicgroup.com	Expo Site
1.16	Sunwin	http://www.sunwinbus.com	Expo Site
1.17	Spectrolab	http://www.spectrolab.com	Belgium Pavilion
1.18	Volvo Group	http://www.volvo.com	Sweden Pavilion
1.19	China FAW Group	http://www.faw.com.cn	Expo Site

UNIDO-ISEC and Jiangsu Modern Low-carbon Technolgy Research Institute (eds.),
The Research Report on Application of Low-carbon Technology in Expo 2010 Shanghai,
DOI 10.1007/978-3-662-44357-6

No.1.1

Company name	Anhui Ankai Automobile Co., Ltd.	Application pavilion	Expo Site
Website	http://www.ankai.com		
Key products and services	Anhui Ankai Automobile Co., Ltd. is a Chinese company engaged in the manufacturing of passenger vehicles and relevant chassis and other auto parts. Anhui Ankai Automobile provides its products under the brand named ANKAI, including sleeping berth passenger vehicles, regular passenger vehicles, urban buses, automobile chassis and automobile spare parts, among others. Anhui Ankai Automobile also offers automobile repair and maintenance services. The Company has three major subsidiaries and one affiliate, which involved in manufacture and sale of passenger vehicles and spare parts		
	The main products are "Kai—Sai Tela" large luxury passenger vehicles, "Ankai" series highway buses, tourist buses, buses, new energy cars (hybrid, electric) and a series of bus chassis, "JAC—Modern" series passenger car, "JAC" series long-distance buses, tour buses, transit buses and light buses, trucks and buses and other auto parts for supporting trucks and buses		
	"AnKai Bao Stone" pure electric passenger travel buses are made with the patented technology, with a "dual power" mode, powered by lithium battery + super capacitor alternatively. This can not only meet high power demand when the vehicle starts, climbs and other conditions, but also extend battery's cycle life		
Patent	Independent Patent		

No. 1.2

Company name	BYD Company Limited	Application pavilion	Expo Site
Website	http://www.bydauto.com.cn		
	http://www.byd.com		
Key products and services	BYD Company Limited was founded in 1995, specializing in two major industries: IT and automobile		
	BYD—a subsidiary of BYD shares directly. Its automotive products include high, medium and low-end series fuel cars, and car mold, auto parts, dual-mode electric vehicles and pure electric cars. Representative models, including F3, F3R, F6, F0, G3, L3, and other traditional high-quality fuel vehicles, S8 sports hard top convertible sports car, SUV, S6 and high-end MPV models M6, as well as the global leader in dual-mode electric F3DM vehicles and pure electric vehicles such as E6		
Patent	Independent patent		

No. 1.3

Company name	Peugeot	Application pavilion	France Pavilion
Website	http://www.peugeot.com		
Key products and services	Peugeot is one of the major brands of PSA Peugeot Citroën Group, Peugeot showcased at the Expo a concept car called BB1. This car's body is only 2.5 m long, but it provides four usable seats. The car's turning radius is only 3.5 m, it can even stop on the sidewalk. The car's side mirror has been replaced by a reverse image, In addition, this car's door is opened to its back, different from the common cars. As a new electric concept car, the car's two electric motors are installed in two rear wheels, with the maximum output power of 15 kW. It takes just 2.8 s from start to accelerating to 31 km/h, The two lithium-ion batteries were installed under the rear seats. The car's mileage can reach 121 km. The overall structure of the body is made of carbon fiber material, The whole weighs less than 600 kg. The technology used includes dual Y-shaped configuration front and rear suspension, and electric power steering. The biggest highlight is the unique roof design, which not only enlarges the headroom of the passengers, solar panel is installed on the roof		
Patent	Independent patent		

No. 1.4

Company name	Chongqing Changan Automobile	Application pavilion	Expo Site
Website	http://www.changan.com.cn		
Key products and services	Chongqing Changan Automobile Co., Ltd. is the core vehicle business affilated to China's Changan Automobile Group Co., Ltd, with an extensive lineup of micro-cars, cars, buses, trucks, SUV, MPV, and other low-middle-grade, wide range and variety of products pedigree, with engine displacement from 0.8 to 2.5 L		
	The new energy vehicles displayed at the Expo include: Changan Zhixiang fuel cell vehicle, Changan pure electric vehicles, hybrid vehicles. It has mastered the core technology for hybrid vehicles, such as the vehicle system integration and calibration match, CAN communication protocol optimization, safety control strategy design, development of diagnostic systems. The most core technology—vehicle control technology, nickel-hydrogen battery management technology		
	6 electric vehicles were put into running at the Expo site, used for Expo transportation		
Patent	Independent patent		

No. 1.5

Company name	Delft University of City Technology	Application pavilion	Netherlands Pavilion
Website	http://www.tudelft.nl		
Key products and services	Delft University of Technology is one of the world's leading Universities of technology, and known as the MIT of Europe. The high-speed solar driven car developed by Delft University of Technology is a single solar-powered car which can speed up to 145 km/h. It has won four consecutive world champions in the World Solar Challenge race held in Australia.Nuna5, covered by 6 m^2 of solar panels, also with a car battery to store excess electricity. The car can still run additional distance when there is insufficient sunlight. When necessary, it can also accelerate the speed to improve its competitiveness in the race		
Patent	Independent patent		

No. 1.6

Company name	Venturi	Application pavilion	Monaco Pavilion
Website	http://www.venturi.fr		
Key products and services	The Venturi Volage is the result of close technological collaboration between Venturi Automobiles and Michelin, presented in Europe during the 2008 Paris Motor Show, then in Monaco Pavilion at Shanghai Expo 2010. This new Venturi has its bodywork made of carbon fibre. The most technical highlight is that each wheel of the car is incorporated with two electric motors, With 4 drive wheels with active suspension. This is the key of "Michelin Active Wheel" technology. Thanks to this innovative design, the Volage can suit different roads and driving method		
	The lightweight chassis of Venturi Volage is evolved from Venturi Fétish and its perfectly mastered weight of 1,075 kilos enable it to attain 150 km/h in less than 5 s		
Patent	Independent patent		

No. 1.7

Company name	Toyota	Application pavilion	Japan Pavilion
Website	http://www.toyota.com		
Key products and services	Toyota is one of the world's top ten automotive companies and Japan's largest automobile company, which was founded in 1933. The early Toyota, Crown, optical crown, crown cars were very famous at the time. Now the Cressida, Lexus luxury cars enjoy great popularity		
	Japan Pavilion exhibited Toyota RAV4 pure electric drive expected to be launched in 2012		
Patent	Independent patent		

Characteristics of six-seat tour car

Technical parameters			
Controller	4 kW		
Battery	48 V 210 Ah		
Motor	4 kW Strong traction motor		
Charger	Computer-controlled intelligent Digital multi-stage charger		
Charging time	8 h (discharge rate of 80 %)		
Performance parameters			
Rated crew (person)	6	Drive type	Rear-wheel drive
Dimensions (mm) (length × width × height)	3900 × 1400 × 1800	Brake Type	The double-pipe of before and after
			Hydraulic drum
Driving range (full load) (km)	80	Front track/Rear track (mm)	1230/1200
Maximum speed (km/h)	29–33	Minimum ground clearance (mm)	205
Minimum turning radius (m)	5.5	Braking distance (m)	5.5
Maximum grade ability (full load)	25 %	Curb weight (kg)	1,080
Wheelbase (mm)	2,730	Vehicle quality (kg)	780
Rated power (kW)	4		

No. 1.8

Company name	Jiangsu Xinri E-Vehicle Co., Ltd.	Application pavilion	Expo Site
Website	http://www.xinri.com		
Key products and services	Jiangsu Xinri E-Vehicle Co., Ltd. is a large private company, specializing in the R&D, production, and distribution of electric vehicles and key spare parts and components		
	The main products: the lithium series, Qingyang series, four-wheel series, 760 series, flying series, load king series, Customized series, FY series, tricycle series, Fenghua series, Yunyan series, Meteor series, the flagship series, Fengya series, God of War series, Fengxing series, wind speed series, etc.		
Patent	Independent patent		

(See No. 1, 1.1, 1.2, 1.3, 1.4, 1.5, 1.6, 1.7 and 1.8)
(See No. 1.9, 1.10, 1.11, 1.12, 1.13, 1.14, 1.15, 1.16, 1.17, 1.18 and 1.19)

No. 1.9

Company name	Yadea Technical Development Co., Ltd.	Application pavilion	Expo Site
Website	http://www.yadea.com.cn		
Key products and services	Jiangsu Yadea Technical Development Co., Ltd. is a large-scale professional and modern manufacturing enterprise of electric vehicles and special vehicles in China		
	Yadea provided the Expo with electric quick cleaning cars, electric three-function sanitation vehicles (ash-bin carrier, road dust absorption and washing), lithium-ion electric road sweeper. By applying the new technology and new materials, Yadea's small electric sanitation devices produce zero emissions, zero pollution, operating in closed system		
Patent	Independent patent		

No. 1.10

Company name	Lifan Group Corporation	Application pavilion	Expo Site
Website	http://www.lifan.com		
Key products and services	Lifan Group corporation was a subsidiary of Chongqing Lifan Holding Co., Ltd, founded in 1992. It is now a large private enterprise focused on development, production, and sale of motorcycles, cars and engines and meanwhile with its services in investment and finance. Lifan Electric Vehicles were specifically developed for the shanghai Expo		
	Lifan provide a total of 30 cars for the Expo, 25 of which were used as a guarding car, and the remaining five were used as VIP shuttles outside the park. Lifan 620 electric cars went on a patrol within the Expo park, In addition to its zero-emission, it has increased the performance of electric cars several times. After charging, the maximum mileage is over 200 km, the maximum speed can reach 120 km/h, the maximum climbing angle is up to 25°, km/h acceleration time in 0–50 is 9 s, and power consumption per hundred kilometers is less than 20 kWh		
Patent	Independent patent		

No. 1.11

Company name	Mercedes-Benz	Application pavilion	Swiss Pavilion
Website	http://www3.mercedes-benz.com		
Key products and services	Mercedes-Benz is one of the global top 10 automobile companies, founded in 1926. It is also the world's well-known manufacturer of big passenger cars and heavy trucks. As the world's most successful luxury car brand, in addition to Mercedes—Benz brand, it also has two other well-known brands: Maybach and Smart		
	The emission of Smart is only 0.9 L. Low engine displacement, to a certain extent, determines its low emission, less than 5 L of fuel consumption per hundred kilometers. Smart environmental approach is not only reflected in the product's environmental performance, but goes through every aspect of the production. Smart factories even have dedicated natural gas thermal stations to ensure the supply of hot water and part of electricity in factories		
	The panels on Smart car body all use recyclable composite materials. Smart maintains its top position among the most environmentally friendly cars with a combustion engine		
Patent	Independent patent		

No. 1.12

Company name	Nanjing NAC Special Purpose Vehicle Co., Ltd.	Application pavilion	Expo Site
Website	http://www.yjzyc.com		
Key products and services	Nanjing NAC Special Purpose Vehicle Co., Ltd. Belongs to Nanjing Auto Group (NAC), specialized in development, production and sales of all kinds of special modified cars. After years of development, it has become China's largest light vehicle modification company covering the most extensive industries. The company has successfully developed and manufactured twelve series of products: engineering trucks, medical vehicles, military vehicles, MPV, commercial vehicles, police vehicles, escort vehicles, promotional vehicles, van (refrigerated truck) vehicles, municipal vehicles, road maintenance machinery, etc., hundreds of special cars and special vehicles		
	The modified vehicles provided for the Expo are ShenChi four seats electric shuttle vehicles, sightseeing vehicles in Expo area—ShenChi 11 seats fuel cell tour buses, VIP reception car—fuel cell cars, Expo VIP reception car outside the Expo—a hybrid cars, Expo Boulevard Line Shuttle—United Nations-funded project fuel cell buses, and taxis outside the Expo—LaCrosse hybrid cars		
Patent	Independent patent		

No. 1.13

Company name	Norway ELBIL NORGE	Application pavilion	Portugal Pavilion
Website	http://www.puremobility.com		
Key products and services	METRO Buddy small pure electric car, its Length is 2,440 mm, Width 1,495 mm, height 1,460 mm, weight 795 kg, which is characterized by angular appearance. For Interior, the high texture of resin components is used. The car is equipped with a motor maximum power of 13 kW, the maximum speed is 80 km/h. Standard models is equipped with lead storage batteries. At full power the continuous driving distance is 50–60 km, with Ni-MH (Ni-MH) rechargeable battery, continued to travel a distance of 80–90 km. In addition, with the lithium-ion rechargeable battery made by United States Valence Technology, the continuous travel distance increases to 120 km. The price of Standard car is €17,000		
Patent	Independent patent		

No. 1.14

Company name	Shanghai Volkswagen	Application pavilion	Expo Site
Website	http://www.csvw.com		
Key products and services	Shanghai Volkswagen Automotive Co., Ltd. (Shanghai Volkswagen), is a Sino-German joint venture founded in 1985. Shanghai Volkswagen focuses on the research and development of new energy, alternative fuel technology and new technology to further enhance the products with low emission levels, By 2010, through it application of TSI, DSG and other cutting-edge technology, it has reduced the vehicle fuel consumption standards to 20 % lower than the current limit. As a result of forward-looking R&D, the hydrogen fuel cell Passat car has low energy consumption, zero-emission. In 2008 during the Beijing Olympic Games it operated as a demonstration of environmentally friendly vehicles and was selected as the Marathon lead car. LPG (liquefied petroleum gas), CNG (compressed natural gas) have gone into mass production		
	The new energy models: Passat fuel cell cars, Shanghai Volkswagen Passat hydrogen fuel cell vehicles		
Patent	Independent patent		

No. 1.15

Company name	SAIC Group	Application pavilion	Expo Site
Website	http://www.saicgroup.com		
Key products and services	SAIC has supplied the World Expo with nearly one thousand pure electric vehicles, super capacitors, fuel cells and hybrid vehicles, including the new-generation "Shanghai" brand fuel cell cars, "Shanghai" pure electric cars, Roewe 750 hybrid cars, LaCrosse hybrid cars, Shenchi four-seat pure electric pavilion cars, Shenchi fuel cell sightseeing cars, the General Equinox, super capacitor buses, hybrid buses, Pure electric buses, and fuel cell buses and electric light buses		
	The new-generation "Shanghai" fuel cell cars are new products SAIC developed specially for the Expo. The performance of the vehicle has reached the world leading level		
	LaCrosse ECO-Hybrid in the first hybrid car among the high-class cars in the domestic; it carries weak hybrid systems of smart ECO engine, Ni-MH battery and permanent magnet AC motor and uses BSG with the function of starting/stopping engine and brake energy recovery. The car can generally save 15 % of the oil and lower 15 % of CO_2 emission, meeting the IV emission standards. 350 LaCrosse ECO-Hybrid cars will be used as the foreign guests reception car in the Expo Garden		
	LaCrosse hybrid cars, Shenchi four-seat pure electric pavilion car launched by SAIC, use high-strength glass-enhanced fiber body with novel outlook. It is equipped with a capacity of 200 Ah lithium batteries, matched with a 3 kW motor with its maximum speed of 35 km/h and a charge driving range of 35 km/h		
	The ShenChi eleven-seat fuel cell tourist car is characterized with high efficiency and low noise, which will be used to carry visitors in the Expo Garden. It adopts the electric hybrid technology programs combining the fuel cell and the power battery with the maximum speed of 40 km/h, 0–100 km/h acceleration of 8 s and a driving range of 75 km		
	SAIC Roewe displayed the Roewe E1 in China Pavilion.The Roewe E1 are A00 single compartment three four-level car with the car body design using "OneBox" concept for the maximum expansion of the vehicle interior space. E1 of the "Zero discharge", is equipped with the safe lithium iron phosphate battery system with the maximum speed of 120 km/h, a top driving range of 135 km and 0–100 km/h acceleration of 16 s. The car has a quick charge function, charging 80 % within 30 min		
Patent	Independent patent		

No. 1.16

Company name	Sunwin	Application pavilion	Expo Site
Website	http://www.sunwinbus.com		
Key products and services	Shanghai Sunwin Bus Corporation (Sunwin) is a Joint Venture invested by SAIC Motor Co., Ltd. (SAIC Motor), Volvo (China) Investment Corp. (VIC) and Volvo Bus Corp. (VBC)		
	Sunwin's ultracapacitor vehicles, fuel cell vehicles, Sunwin pure electric city buses, hybrid buses lead in the domestic market. Sunwin super capacitor car use supercapacitor produced by Aowei Technology Development Co., Ltd. Sunwin pure electric city bus use manganese lithium-ion batteries of CITIC Guoan MGL (Citic Guoan MGL). UN fuel cell city bus uses Shanghai Sunwin Bus Body, equipped with fuel cell stack developed by the Canadian Ballard Power company. Sunwin hybrid buses is equipped with 5.88 L 185 kW diesel engines, 576 V, 60Ah lithium-ion iron phosphate battery, 110 kW AC induction drive motor and hybrid system consisting of 70 kW, ISG Integrated Starter/generator, with a maximum speed of up to 80 km/h		
Patent	Independent patent		

No. 1.17

Company name	Spectrolab	Application pavilion	Belgium Pavilion
Website	http://www.spectrolab.com		
Key products and services	Spectrolab, headquartered in Sylmar, California, is the world's leading manufacturer of multi-junction solar cells, search light, and solar simulators		
	At the Shanghai Expo, Spectrolab showcased a solar car called umicar. It is high-efficient and lightweight		
	Umicar specifications: 4.30 m long, 1.80 m wide and 1.40 m high, the whole car efficiency: >30 %		
	Air Power: The umicar infinity improve the air resistance by 25 %; weight <170 kg; top speed of 140 km/h, cruising speed: 100 km/h		
Patent	Independent patent		

No. 1.18

Company name	Volvo Group	Application pavilion	Sweden Pavilion
Website	http://www.volvo.com		
Key products and services	Volvo Group was founded in 1927 and headquartered in Gothenburg, Sweden, Volvo classic vehicles inlcude cars (S series), business wagon (V series), SUV/sports and leisure vehicles (XC Series) and convertible/Coupe (C series). In Sweden Pavilion of the Expo, it displayed the Volvo hybrid city bus and the Volvo hybrid garbage truck. Volvo hybrid city buses include Volvo7700Hybrid and VolvoB5Lhybrid double-deck bus, The bus features the parallel two engines: diesel engines and electric motors to work together. It is a hybrid in that the bus is powered by a combination of an electric motor and a diesel engine, with the system being parallel in that it can be powered by either source independently, or in combination drawing on both power sources at the same time. The fuel consumption and CO_2 emission can reduce by 30 %. Compared with the conventional diesel buses, the nitrogen oxides and particulate matter emissions reduce by 40–50 %. Volvo FE hybrid garbage truck saves the fuel by 20 % when treating the wastes, 15–20 % when in urban road transportation		
Patent	Independent patent		

No. 1.19

Company name	FAW Group	Application pavilion	Expo Site
Website	http://www.faw.com.cn/		
Key products and services	FAW Group is one of the leading manufacturers in China with 50 years history		
	FAW Pentium hydrogen fuel cell cars displayed in the Expo Park is one of the research results of the national key project "Energy saving and new energy vehicles", As an item of national "863 Program", and the "energy-saving and new energy vehicles", R&D project, this car meets the international standards of environmentally friendly vehicles, in safety, reliability, comfort, etc. which is a key product of FAW Pentium independent R&D		
	FAW has independent R&D capability in core technology of commercial vehicles, passenger cars, mini and light cars, engines and automotive electronics, etc.		
Patent	Independent patent		

MIX
Papier aus verantwortungsvollen Quellen
Paper from responsible sources
FSC® C105338

If you have any concerns about our products,
you can contact us on
ProductSafety@springernature.com

In case Publisher is established outside the EU,
the EU authorized representative is:
Springer Nature Customer Service Center GmbH
Europaplatz 3, 69115 Heidelberg, Germany

Printed by Libri Plureos GmbH
in Hamburg, Germany